Feasible Living

Dealing with Ecological Anxiety While Adapting to Our Changing World

Editor: Linda Parsons
Cover Design: SelfPubBookCovers.com/Abek
Book design: Ken Kroes
ISBN: 978-0-9958470-6-4
Publisher: 1779671 Alberta Inc.

Ken Kroes

www.feasibleplanet.com

DEDICATION

I would like to dedicate this novel to my wife Linda for her continuous encouragement and my children, Megan, Jarid, James, and Julia, who are the reason for writing it.

ACKNOWLEDGMENTS

I would like to acknowledge the following people for their efforts and invaluable feedback that I received while writing this book.

Linda Parsons for her tireless editing of multiple revisions and her input on several key sections.

Richard Schwindt for having the courage to go through an initial draft and providing pragmatic observations.

Herb Simmens for his detailed input and vocabulary suggestions that are also published in his recent book, "A Climate Vocabulary of the Future".

Laura Wilkinson for her meticulous review and suggestions.

Contents

Introduction

At times, it seems that things on our planet are out of control. Headlines on topics like climate change, pollution, mass extinction, government debt, and robots stealing our jobs are common and whip up a wide range of emotions.

Fear of the future, **rage** at a lack of consolidated effort to rectify the problems, **sadness** for damage that is being done, and **helplessness** of not being able to do anything are some of the feelings many of us have. These types of feelings lead to heightened **anxiety** within each of us, including, and perhaps most importantly, within our children.

The concern regarding the relationship of living organisms **to one another and their physical surroundings** has a specific name: ecological anxiety. **Ecological Anxiety Disorder** (EAD) occurs when this concern elevates to a point where it starts to impact our daily lives.

Should we be concerned? After all, there have been many dire predictions in the past that did not come to pass. Are things different this time? In short, the answer is yes.

The difference between then and now is that we are reaching a **peak population** on this planet. The next several decades are special, indeed unique in the history of mankind, in that most forecasts predict our population will start to level off. A time that you, your children or grandchildren will see. Unlike our ancestors who saw unlimited growth, it appears we will reach an era where our population, and eventually our demand for goods and services, will mature.

Many of the reasons for our population peaking are related to the drivers of our ecological anxiety. Availability of resources, pollution, economic, and social trends are limiting our ability to further expand. In fact, it appears that we may have already overshot the limits of what our planet and society can support.

Ecological anxiety is a serious issue and can lead to a variety of physical and mental health issues including depression and a general feeling of hopelessness. It can also drive us to a behavior that is not in our best interest: **doing nothing**. The problems that our planet is facing are real. They are impacting us right now and will only get more extreme as time goes on. By not acting **now** to safeguard yourself, you will only make the situation worse for you and your loved ones.

What can be done to lower your ecological anxiety? What actions should you consider taking right now? These are the questions that this book will hopefully answer for you.

A big part of this anxiety is the **fear** that we feel about the problems; the not knowing about what will happen. To reduce this fear, the first section of this book will briefly cover the major issues facing us today.

You may think that reading about the issues will only increase your anxiety, but fear is the emotion that we have when we do not know what is coming at us. In other words, **fear is all about the unknown**. Knowing about the problems will allow you to act. You may still dread the outcome, but you know what will happen.

The second part of the book is all about you. How you may be impacted by the eco-problems that the world is facing, and what you can do to reduce the impact on you and your loved ones. Note that this section **is not focused on saving the planet.** For example, it does not discuss actions you can take to lower air pollution but instead has suggestions on how you can minimize the impact of air pollution on *your* health.

If part of your ecological anxiety therapy is to work towards saving the planet, my recently published book, *Feasible Planet - A Guide to More Sustainable Living,* may be for you. This book deals mainly with actions that you and I can take now to live more sustainably, thus buying more time for governments and businesses to get things right. The premise of *Feasible Planet* is that we need not agree on what is causing our issues or on the long-term solutions, just that we are not certain how things will turn out and that the prudent decision is to act.

One last note, while this book does get into some detail, it is meant to be an introduction to a wide range of topics. I have included many references for further reading on subjects that may interest you and I encourage you to explore them.

> Inaction breeds doubt and fear.
> Action breeds confidence and courage.
> Dale Carnegie

Key Issues

The first section of this book will briefly go through twenty key issues (all trending for the worse) that are impacting us now and will shape how we live in the coming decades. None of these trends will abruptly change the world, but each will approach a boiling point within the next few decades; a time that you or certainly your children will live through.

There is no effort in these brief introductions to convince you that the trend is real or not. Barring a global pandemic or world war, these trends and their consequences are pretty much locked in for the next few decades. If you choose **not** to believe that one of the trends is 'real', that is your choice, but in my opinion, the facts firmly indicate that they are.

If you have an interest in helping the planet by living more sustainably, each of the key issues has a ***Feasible Planet Index Rating*** attached. This is my **subjective** take on your level of control over your contribution to each issue combined with how much impact your action could have on the subject.

The rating scale is from 1 to 9, where 1 indicates little impact and 9 meaning you have a great deal of control over your contribution and it has a significant impact. For example, your actions determine the size of your carbon footprint, so I have given that a rating of 9. However, if we look at the issue of increasing world population and migration, it is mostly out of your control, so I have given this a 1 rating.

To get more details on how you can help, my book, "*Feasible Planet – A guide to more sustainable living*", is packed with suggestions.

Water

Water Pollution

Feasible Planet Index

6

How much can you help?

1 = Low impact
9 = High impact

Do you think the water from your kitchen faucet is just water? The answer is nearly certain to be no. There are a wide range of pollutants that can be in your water regardless if you live in the city or out in the country.

Lead from aging water infrastructure, arsenic leaching into groundwater supplies, industrial and agricultural waste that is fed into rivers, and fracking are just some things that may contaminate your drinking water supply.

To keep the water safe from bacteria, viruses, and protozoans, most cities add chlorine to the water. While this makes the water safer to drink, it has side effects such as dry skin and causing rashes. Other less known side effects are the potential by-products of the chlorine chemical reaction, Trihalomethanes (THMs) and Halo-acetic acids (HAAs). Both of which have been linked to a variety of diseases including bladder and kidney failure as well as various forms of cancer [1]. As more contamination enters our water supply, most cities are using an increasing amount of chlorine to disinfect it, leading to a higher risk of exposure to these chemicals.

> The Environmental Protection Agency (EPA) has categorized THMs as a group B carcinogen [236].

One of the more recent chemicals of concern in our water goes by the acronym PFAS (Per- and polyfluoroalkyl substances) which has been used for decades in common items such as cookware, stain repellant, and even some types of pizza boxes. The impact of this group of chemicals on humans is not entirely understood but studies are negatively linking it to issues with cholesterol, immune systems, reproduction, liver, kidneys, and other organs. A recent study found that **110 million Americans could have PFAS contaminated drinking water** [2].

Despite the best efforts to keep our water safe, sometimes things do not end well such as the case of lead in the residential water of Flint Michigan. There are also millions of homes in North America that currently get water from systems that violate regulations [3].

Chemicals are not the only thing to worry about in our drinking supply – microplastics are found in nearly all water systems. These microplastics can carry bacteria and other toxic chemicals into our bodies and a recent study has found that most tap water contains them [4]. All of this may lead you to want to drink only bottled water but not so fast! A recent World Health Organization study found that over 90% of bottled water, including many popular brands, also contain microplastics [5].

Every living thing on the planet needs water, including the ones we eat. Several forms of pollution like arsenic, mercury, and PFAS's do not biodegrade very well and accumulate in the food we eat. A good example of this is the EPA fish advisories that give limits on the maximum number of freshwater fish that can be eaten to avoid health concerns.

Rivers, lakes, and oceans face even more pollution issues than your tap water does. Run-off from agricultural land encourages dead zones in our oceans. Some cities pump raw sewage into the ocean while others have outdated water treatment facilities that overflow during storms. This is all on top of the piles of plastic waste that find their way to both the top of the ocean surface and the sea bed.

Estimates show that, by weight, there will be more plastic than fish in the ocean by the year 2050 [209].

Our polluted water also impacts our recreational activities. Each year beaches are closed because the water is not safe to swim in because of pollution that we have put into the water. In the United States, **one in ten beaches are classified as unsafe** for swimming for at least a part of the year (6).

With the lack of funding to fix our aging water supply systems, increased consumption, population, and ongoing use of plastics, water pollution will be a problem that will continue to get worse for some time to come. Since water is such a broad topic, further points related to pollution in it are covered a bit later, such as in the section "Nitrogen Cycle".

Fresh Water Shortage

Feasible Planet Index

7

How much can you help?
1 = Low impact
9 = High impact

How secure is your water supply? While the last section went through the pollution that is in our water, there are nearly a billion people on the planet that lack even basic drinking water services [7] and four billion experience shortages for at least one month of each year [8].

Regardless of where you live, many water sources are in a stressed condition. From depleting aquifers, shrinking glaciers, rivers that no longer make it to the ocean, and record-low water levels in critical reservoirs, many of us could soon have our usage tightly regulated or see our taps run dry. Overdramatic? Look at these examples and then be the judge:

- **Yemen**, a mainly agriculture-based country of 27 million people, is drawing down its aquifers at such a rapid rate that they will be depleted in a few years [9]. Yemen is not the only place where aquifers are under stress. Globally, 2 billion people rely on aquifers for their primary water source and two-thirds of the major aquifers are being drawn down faster than they can replenish [10].
- While on the topic of aquifers, those that are close to the coast, like in **California**, are being drawn down so drastically that saltwater from the ocean is seeping in, ruining the aquifer for future irrigation and human consumption. Water contamination from the oil and gas industries are also a threat with their activities compromising aquifers in several states [11].
- In the United States, **Glacier National Park** was established in 1910 and had 150 glaciers, now it has 30 and most of these have shrunken by two-thirds. It is estimated that all the glaciers in this park will be gone by 2050 [12].

- The glaciers in our mountain ranges provide fresh water that is used by everyone. As these glaciers get smaller, there comes a point where the runoff also decreases. This is referred to as a **tipping point** and it is estimated that this tipping point has already been reached in roughly half of the glaciers around the world (13).
- Several major rivers around the world no longer make it to the ocean, especially in the dry season. Examples include the **Colorado River in North America, the Yellow River in China, the Nile in the Middle East, and the Ganges in Southern Asia (14).**
- Due to a variety of factors, several key water reservoirs such as **Lake Mead** in the United States are very near all-time low levels.
- In early 2018, **Cape Town** in South Africa was within 100 days of running out of water. Thankfully, enough rain came to avert a crisis (at least for that year). Other major cities that are facing water shortages include **Mexico City, Bangalore, Tokyo, London, and Miami** (15).

Why are we running into these shortages? Surprisingly, it is **not because we are using more water**, at least in industrialized countries. Water consumption has actually decreased over the last several decades (16) because of improvements in electricity generation facilities, which are major consumers of water in industrialized nations.

This leaves the primary issue for our water shortages to be blamed on prolonged changing weather patterns (i.e. climate change) with more heat and less precipitation in certain areas. Another major factor is our increasing population and the location of these populations relative to where fresh water is.

The prognosis of this situation is not good. There is a big push right now on generating more electricity and a fair bit of this is going to natural gas generation stations, which use water. There is also the increasing global temperatures and the forecast of another 2 billion or so people on the planet within the next three decades.

Water is life. If an area of the globe becomes severely stressed, people will migrate, wars could erupt, and there will be an additional strain on other areas of the world for the increased demand in food production.

Our Changing Oceans

Feasible Planet Index

6

How much can you help?

1 = Low impact
9 = High impact

Regardless if you live on the coast or inland, our oceans are probably the biggest source of planetary change that is impacting you.

This is because our oceans are absorbing much of the extra energy that our planet is retaining due to increasing greenhouse gas levels in our atmosphere. Recent studies estimate that 90% of warming that has occurred on Earth in the last 50 years has occurred in our oceans (17).

This change in temperature combined with decreasing salinity levels (18) are the major causes for the changes in our ocean currents and for the recent **changes in the behavior of the jet stream** (19) **which influences weather patterns around the globe**.

The current theories on the many extreme weather events that we are experiencing recnetly are not just that the planet is getting hotter, but the weather is not moving as quickly as it once did. A heatwave, cold front, or major precipitation event that would normally last a day or two, now may stay in place for a week or longer.

Beyond the weather impact, one of the more publicized predictions from our rapidly changing planet is our **rising ocean levels** due both to ice melt and thermal expansion. A 2018 study from NASA is predicting an increase in level of about two feet by the year 2100 (20). While this may not seem like very much, it can greatly impact the damage done by storms and has already been attributed to decreasing real estate value on coastlines (21).

The increasing level of carbon dioxide (CO_2) in our atmosphere is also resulting in increased levels of ocean acidity with the surface waters being 30% more acidic now than at the start of the Industrial

Revolution [22]. This increased acidity is having a major impact on marine life such as shellfish and corals [23] [24].

If part of your diet or business relies on the ocean, that is being impacted too. Decreasing fish populations, dying coral reefs, dead zones, changes in plankton life cycles, loss of kelp forests, and toxic algae blooms are all **current** issues that we are facing due to pollution, rising temperatures, overfishing, and increasing ocean acidity.

There is no current unified effort to curb any of the factors that are driving the rapid changes within our oceans. In fact, the rising global population will most likely result in additional strain which will see worsening conditions in our oceans over the upcoming few decades.

Air

Air Pollution

Feasible Planet Index

7

How much can you help?

1 = Low impact
9 = High impact

We should not be able to see, smell or taste the air, but this is an all too common occurrence in our daily lives. Most of you have probably seen pictures of extreme smog that exists in large cities. Pictures that show nearly everyone walking around with air masks covering their faces and with a visibility of a few miles or less. While these are extreme cases, most of us inhale air pollutants every day that are generated from various sources such as cars, planes, factories, forest fires, and power plants.

Key pollutants are sulfur dioxide, nitrous oxide, volatile organic compounds (VOC), carbon monoxide, ammonia, and particulates. Carbon dioxide and methane are not generally considered pollutants and are discussed in the next section. Ozone can be treated both as a pollutant and a greenhouse gas, and this also has a special section devoted to it a bit later.

The good news is that several of the bad air pollutants like sulfur dioxide and nitrous oxide are on the decline through much of the developed world. However, they are still at dangerous levels in most developing nations. The World Health Organization estimates that air pollution causes about 7 million premature deaths worldwide each year [(25)].

The bad news is that there are some trends that are increasing air pollution. More air travel, the higher demand on power plants, and increased risk of forest fires are a few examples. As I write this chapter, my community has been under heavy smoke from the forest fires on the west coast of North America for the last two months, causing respiratory difficulties in the young and elderly.

Greenhouse Gases

Feasible Planet Index

9

How much can you help?
1 = Low impact
9 = High impact

We hear an awful lot about greenhouse gases (GHGs), climate change, and global warming, but what exactly does it mean and what is the big deal?

The simple version is that greenhouse gases, like carbon dioxide (CO_2) and methane (CH_4), act as a blanket and helps to keep the planet warm. Some of these gases are created naturally, like when leaves decompose, and some are made through actions we take, like burning fossil fuels such as gasoline and coal.

Over the last few hundred years, we have burned a great deal of fossil fuels. This has resulted in more greenhouse gases in our atmosphere and, hence, our planet, overall, is getting a bit warmer. This is where the term, "**Global Warming**" comes from.

This heating of the Earth is having various impacts such as warmer oceans and less reflective ice at the poles. This, in turn, is impacting our normal weather, hence the term "**climate change**". **Yes, the climate has always changed, but never at this pace and certainly never with 7-9 billion people relying on it**.

There are times when I think that keeping up with the developments on this climate chaos is more suspenseful than reading a John Grisham mystery novel. The multiple plots, surprises, hope, despair, conspiracy, and suspenseful endings can keep me up reading until well into the wee hours of the morning. This addiction has taught me that the issues are intertwined and complicated, making precise predictions difficult while giving plenty of fodder for climate deniers to twist around.

Given all this, it is of little wonder that people who do not make climate science their career can get mixed signals on the level of impact our greenhouse gasses have on the planet. But, as I mentioned in the introduction, this section will not try to convince you on "if" climate change is occurring or "if" we are responsible for it. There is a staggering amount of evidence out there in support of it and I have included a few good references at the end of each section in this book for further reading.

Some impacts from increased greenhouse gas levels have been discussed in earlier chapters such as our freshwater shortage, warmer oceans, rising oceans, and the changing climate. Other impacts that may not be so obvious include:

- Nuclear plants shutting down due to intake cooling water being too hot for safe operation of the plant.
- Fish dying or salmon laying eggs prematurely because of high river temps, toxic algae blooms, etc.
- Shifting climate zones. Expanding tropic regions, the expansion of the Sahara Desert, expanding western North America arid region, eastern movement of "tornado alley" in the U.S., shifting plant hardiness zones, and permafrost lines moving farther north [(26)].
- The extra heat contained in the ocean is increasing the strength of hurricanes which, when combined with increased sea levels, can result in flooding at record levels as witnessed by hurricanes Florence, Harvey, and Idai.
- Increased plant growth. With more carbon dioxide and warmer temperatures, there has been an explosion of green growth. According to NASA, a quarter to half of the Earth's vegetated lands have shown significant greening over the last 35 years. This is perhaps a good thing, but it is also impacting the water cycle in many regions [(27)].

Even if we were to stop **all** emissions today, the impact of the emissions we have already put in the atmosphere will last for centuries and we are a long way from cutting off our emissions. In fact, we are doing the exact opposite with increasing oil and natural gas demand! There are other sources for greenhouse gases, including defrosting permafrost, wetlands, and deforestation; all of which are also trending in the wrong direction.

Ozone

Feasible Planet Index

5

How much can you help?
1 = Low impact
9 = High impact

Ozone (o_3) is a bit like tequila in that there are good and bad kinds. Just like "top shelf" tequila is usually the good stuff, the ozone in our upper atmosphere blocks and protects us from the Sun's ultraviolet radiation. Bad tequila, the stuff not made from 100% blue agave, is usually on the lower shelves and tends to give a nasty hangover. The bad ozone is also on the lower shelf, in our case the lower atmosphere, and is not very good for us either.

Lower atmosphere ozone or ground-level ozone is sometimes referred to as smog and is formed when hydrocarbons and nitrogen oxide combine in sunlight and heat. If inhaled, ozone can constrict your airways, making it harder to breathe. This is especially a problem in younger and older populations and those that already have breathing difficulties.

Smog can last in the air for days and, if it drifts over farmland, can reduce crop yields, especially in rice, maize, soybeans, and wheat fields by about 10% [(28)].

With the trend for higher temperatures and little near-term action to reduce fossil fuel consumption, it appears that more smog is in our future for at least the next few decades and will continue to negatively impact our health and food supply.

The upper atmospheric ozone has had challenges as well. Most of you probably know that chlorofluorocarbons (CFCs) are destructive to our ozone layer. This was discovered as early as 1974, yet it took until the year 1990 for world powers to set up regulations to ban CFCs in developed countries by the year 2000 and developing nations by 2010.

The good news is that the ozone in the upper atmosphere can replenish itself quickly. Recent reports show that because of global CFC bans, by 2050 the ozone in the upper atmosphere should be back to the same level as it was in 1970. Things are still not perfect though since there are still several chemicals used in industries that impact the upper atmosphere ozone layer. Ongoing meetings are occurring to come up with new regulations to control these.

Land

Soil Pollution and Degradation

Feasible Planet Index

3

How much can you help?

1 = Low impact

9 = High impact

With most people in industrialized countries living in urban centers, their only exposure to soil is the grass-covered areas in parks, boulevards, and lawns. There may be an occasional flower bed or garden, but that is about it. Very little attention is given to the productive soil that we and the rest of nature rely on to support our ecosystem and food supply. The same productive soil that is under attack on several fronts.

Pollution of our soil is a growing concern and directly impacts every living being on the planet. Industrial waste, wars, intense agriculture, garbage disposal, and mining are some of the bigger contributors to soil contamination. How bad is it? Here are a few examples that concern me:

- Rice grown in the United States, particularly in the south-central region, has alarming levels of arsenic. This arsenic comes from pesticides that were applied to cotton fields before they were switched over to growing rice [29].
- China is reporting that 19 percent of its agricultural soils are polluted and there are about 1,300 heavily-polluted sites in the United States [30].
- There has been much coverage on microplastics in our water, but did you realize that even more plastic is in our soil? These microplastics can promote diseases in our food and even impact the behavior of essential and hard-working worms [31] [32]!
- A "small one", but one of my personal favorites. If you live in Canada, ***until the end of time***, taxpayers will pay to

freeze the ground near Yellowknife to prevent over 200,000 tonnes of arsenic (that we put there) from seeping into Northern Canada's waterways. The arsenic is from gold mining operations and one-fifth of one gram of arsenic is a lethal dose.

Another critical issue that is being faced is soil erosion. With many farmers growing the same single crop year after year, water and wind are given ample chances to remove the topsoil from our fields. Other factors include overgrazing, droughts, and deforestation.

The United Nations Food and Agricultural Organization (FAO) released a recent study that, on average, there are about 60 more harvests left worldwide before the topsoil is severely depleted. Now, this is an average. In the UK and North America, the number is higher, with over 100 harvests. But in regions of high population, such as India, China, and parts of Africa, the number is less than 60 [33] [34]. Where are these billions of people going to get their food from in a few decades?

> The nation that destroys its soil destroys itself.
>
> Franklin D. Roosevelt

The chemical makeup of our productive soil is also showing worrisome trends. Organic matter, which can take decades to rebuild, is on the decline in many areas due to lack of crop rotation and persistent use of tilled-based crops [35]. Much of our soil is also becoming more acidic, due, in part, to nitrogen-based fertilizer usage [36].

This topic does not get much attention but consider this: Even if we solve global problems of pollution and climate change, these trends still exist and will be the trigger point for mass migration and political instability.

Deforestation

Feasible Planet Index

4

How much can you help?
1 = Low impact
9 = High impact

The importance of our forests to the sustainability of this planet cannot be overstated. Not only do they provide an environment for wildlife biodiversity, but they are also absorbing a great deal of carbon dioxide. The issue here is that the size of our forests is decreasing mainly because of mining, an unprecedented demand for wood products, aggressive agricultural expansion, changing climate, and forest fires.

Though reports were indicating that the rate of deforestation is slowing down, recent studies that use satellite images, have shown that the **rate of deforestation has actually increased** over the last couple of decades to the point of **currently destroying the size of a football field every second** [37]. Most of the deforestation occurs in the tropics. If drastic action is not taken now to curb agricultural growth, there is a good chance that most of the rainforests in this region will be gone in about 100 years [38].

The other large forest area on the planet is called the Boreal Forest (also referred to as the Taiga or snow forest). This forest is in the northern part of the planet and makes up about one half of the total forest on the planet [39]. This massive forest is now in danger because of changing climate patterns. Risks include forest fires, drought, and increased insect infestation due to warmer weather. These factors will turn this forest region into more of a savanna region within the next several decades [40] [41].

The Boreal Forest contains a great deal of carbon and can absorb even more. Losing most or even part of this forest will make keeping climate change under control much harder.

Nitrogen Cycle

Feasible Planet Index

2

How much can you help?

1 = Low impact
9 = High impact

Though there is plenty of discussion on the carbon cycle and the related impact on climate change, there is another cycle that humans disrupt that is equally important – the nitrogen cycle.

The simple description of this cycle is that nitrogen rotates through the atmosphere, land, and oceans while changing into various chemical forms. Nitrogen is critically important to all life on this planet by way of cell development and nutrients in both plants and animals.

Humans are impacting this cycle mainly by using fossil fuels, burning of biomass, and the creation and application of fertilizers. From an agricultural standpoint, much more nitrogen-based fertilizer is typically applied than is required by the crops. Excess nitrogen leaches into the soil and waterways. Once in the water, excess nitrogen can lead to algae blooms and to greatly reduced oxygen levels in rivers, lakes, and the oceans. Other impacts include acidification of freshwater and changes to the biodiversity of ecosystems.

Though tightly regulated in some regions of the world, the burning of fossil fuels does emit a dangerous greenhouse gas called nitrous oxide. This gas is about 300 times more powerful than carbon dioxide as a greenhouse gas and stays in the atmosphere for at least 100 years [42]. It is estimated that 6% of greenhouse gas production in the United States in 2017 is from nitrous oxide [43].

Not only is nitrous oxide a greenhouse gas, but it can also chemically react to produce ground-level ozone and smog. The negative health impacts of this were discussed just a few pages ago in the "ozone" topic. To top that off, some nitrous oxide makes it to the upper atmosphere where it chemically reacts to destroy the helpful upper ozone layer.

One of the most detrimental impacts we are imposing on the nitrogen cycle is the amount of "reactive" nitrogen that we are putting into the global nitrogen cycle. In fact, humans are now the biggest source of "reactive" nitrogen on the planet [44]! This issue is critical. This excess "reactive" nitrogen is the major contributor to biogeochemical pollution; one of the four major tipping points that we are approaching or exceeding that will lead to "irreversible and abrupt environmental change" [45].

The science behind the nitrogen cycle is not as well understood as the carbon cycle, but recent estimates show that to avoid serious biodiversity loss, **we need to cut the amount of reactive nitrogen used by half by 2050, just 30 years from now**.

Mineral Shortages

Feasible Planet Index

2

How much can you help?

1 = Low impact

9 = High impact

Everything we consume is made up of natural resources that we take from our planet. Many of these resources are abundant like iron, silicon, and aluminum, but several such as phosphorus, lithium, and some rare earth minerals are in limited supply.

Now, we are not going to run out of these materials overnight. As some get scarcer, alternatives can be found. But for others, like phosphorus, where there is no practical alternative, the price of the material will go up. This results in making some products more expensive to produce and may lead to more robust exploration to find new sources of raw materials, such as ocean dredging or mining in wildlife reserves.

Some examples are given below:

- Though very hard to estimate, several studies are pegging peak copper, the time when peak production is reached, to be somewhere between 2045 and 2070. Through recycling and the use of alternatives, we may be able to extend this date, but it will come at a price, one that you and I will have to pay.
- Phosphorus is necessary for all life on the planet and it is a key ingredient in fertilizers. According to the United States Geological Services (USGS), if we look at worldwide reserves and demand, we will run out of phosphorus in 250-300 years [46]. Of course, things will get worse long before that as we need fertilizers for our crops to feed the growing population on the planet. Another nasty point here is that unlike copper, **there is no replacement or alternative for phosphorus.**
- To make green technologies such as energy-saving light bulbs or magnets for wind turbines, rare metals such as

neodymium, terbium, indium, and iridium are needed. Current supply sources for these metals are not adequate for the anticipated ramp-up in demand for green products. China currently is the source for about 90% of these rare earth metals and has indicated that if demand continues to grow, their mines will run dry within the next two decades (47) and as such, is now limiting exports.

Shortages that build up over the next few decades will lead to higher prices for all of us and, worst case, political tensions between countries. One does not have to look too far for an example of this either when you see the number of lives lost, and money spent on wars in the middle east that were directly related to oil supply.

Biodiversity Loss

Our Changing Ecosystems

Ecosystems on this planet have been changing since they first formed. Because of mutations, external events, and the natural ebb and flow of life within an ecosystem, some species thrive and others die, but the ecosystem usually stays viable since this happens very gradually, and a healthy ecosystem has many forms of life within it.

A sustainable ecosystem consists of plants and animals that work together to keep things in harmony. The loss of **any** species will disrupt this harmony but, if there are many forms of life, other species can make up for the loss. The more variety of living things or **biodiversity** that exist in an ecosystem, the more **resilient** it becomes.

There have been a few times during our planet's history where changes to the ecosystems were not so gradual. These times were triggered by major events like asteroid strikes and large volcanic eruptions which resulted in rapid depletion of many forms of life. So many, in fact, that they nearly caused entire ecosystems to collapse. These types of events are referred to as mass extinction events [(48)].

Mass Extinction Event	Millions of years ago	Estimated % loss of species
Ordovician-Silurian	450	70
Late Devonian	375	70
Permian-Triassic	252	90-96
Triassic-Jurassic	201	70-75
Cretaceous	66	75

With these mass extinction events, a small amount of life survived and, after millions of years, the ecosystems were restored. Which species survive a mass extinction event or even a less dramatic ecosystem decline depends entirely on **how fast the species can adapt to the changes**.

You should be able to guess where this is leading. The impacts of climate change, excess reactive nitrogen, pollution, and other non-sustainable trends is resulting in **biodiversity loss in our ecosystem <u>right now</u>.**

So far, with the aid of technology, humankind has been able to adapt quickly to changes in our ecosystem. Pesticides, fertilizers, better equipment, and irrigation are some advancements that the industrial food complex has used to provide food for our ballooning population.

But it is not all about technology, nature continues to provide many services that are critical to feeding us. Services that are difficult for technology to overcome, such as pollination, soil enrichment, pollution control, water management, and supplying non-farmed foods like seafood. As biodiversity losses mount, natural services that we rely on will be stressed and so will our ability to ensure that everyone will have enough food.

The impact on us from biodiversity loss is not limited to our food. We rely on other living things in our ecosystems to provide us with clean air, water, jobs, and even medicines. For instance, 40% of all prescription drugs and 70% of newly-developed medicines are derived, at least in part, from plants (49) (50).

If you thrive on pessimism, you will be delighted to know that as we lose biodiversity, we will be making some other trends, like pollution, even worse, creating a nasty feedback cycle.

You may be able to see some of this biodiversity loss around you, but there is a good chance that you will not. **With globalization and urbanization, most of us have very little knowledge of where our food comes from, how nature is involved in the things we consume or the status of the related ecosystems**. But just because we cannot directly see the population decline of plants, animals, and insects does not mean that it is not happening nor that we are not impacted.

Biodiversity Indicators

Biodiversity is complicated. To help measure it, various groups have developed a few key indices, called bio-indicators, that help us keep a pulse on where things are at.

IUCN Red List Index

The International Union of Conservation of Nature's Red List of Threatened Species was started in 1964 and is a list of animals, plants, and fungi along with a category of the level of risk each has for extinction.

As can be imagined, this is a gigantic undertaking. The sheer volume of species and the work required to evaluate and then continue to monitor is staggering. However, the data that is obtained is a vital indicator of the health of our ecosystems.

The current list has **roughly 26,000 species that are threatened with extinction** (51). Currently, corals are moving to extinction most rapidly while amphibians are the most threatened group. Though the data is limited and only goes back a few decades, obvious downward trends can be seen in nearly all the categories.

More information at:

www.icunredlist.org

Living Planet Index

The Living Planet Index (LPI) is a different indicator that we can use to judge biodiversity loss. Unlike the Red List, the LPI measures the population of only vertebrate species. Currently, the LPI tracks just over 4,200 species (52), including mammals, birds, reptiles, and fish.

The most recent report, published in 2018, shows that of all the species monitored, **on average, animal populations were reduced by well over half between 1970 and 2014** (53).

More information at:

www.livingplanetindex.org

Global Wild Bird Index

The population of wild birds is a good indicator of the overall health of an ecosystem. The Wild Bird Index (WBI) is maintained by the Biodiversity Indicators Partnership and measures the population of a suite of representative wild birds.

The most recent report from this group shows a **decline of about 40% in specialized birds and 20% in overall bird populations monitored between 1974 to the year 2014.**

More information at:

www.bipindicators.net/indicators/wild-bird-index

6th Mass Extinction

Many scientists are now convinced that we are at the beginning of our planet's 6th mass extinction. An event that will see a radical change to our planet and threatens much of the human population.

It does not take much imagination to see why the scientific community is thinking this way given trends shown in the biodiversity indices. To clinch the deal here are some key headline points from the May 2019 summary report from the Intergovernmental Science-Policy Platform on Biodiversity and Ecosystem Services (IPBES). This is based on 15,000 scientific and government sources and compiled from 145 expert authors from 50 countries (54).

- Currently one million species are at risk of extinction (roughly 25% of animal and plant species are being threatened).
- Human activity has resulted in the severe alteration of over 75% of Earth's land area.
- Remote tropical forests are nearly silent as the insects have vanished. Overall, it is estimated that 40% of insect populations are on the decline.
- There are 400 dead zones found in our oceans and in parts of the ocean, little life remains except for green slime.
- Overall, wetlands are down to 15% of what they were 300 years ago.

What the future holds in store is a matter of debate and greatly depends on how active we become at saving and restoring natural areas of our environment. At the current rate, we will probably not have massive ecosystem collapses in the next decade or two, but beyond that is unknown and scary territory.

A glimpse into the future of our world can be seen in the current and short-term impacts we are already experiencing, which are discussed in the next section.

Current and Short-Term Impacts

Feasible Planet Index

3

How much can you help?

1 = Low impact
9 = High impact

Monoculture Farming

With the challenge of feeding billions of people, the industrial food complex has focused on raising just a dozen or so types of crops such as wheat, corn, rice, and soybeans. Having fewer types of crops makes the planting, harvesting, and processing easier, but there is an inherent risk with this approach because even though there are several major varieties of each crop, there is little genetic diversity. This can lead to quick and widespread crop loss due to disease. Some examples are below:

A fungus with the common name, **Stem Rust**, hit cereal crops hard in the 1950s. Scientists then developed strains of wheat and barley that were resistant to it, but in 1999, a new strain of this fungus appeared, and most crops are not resistant to it. Stem rust is currently spreading across Africa, Asia, and Europe as scientists work to develop new strains of crops resistant to this mutated disease. The last few outbreaks of this disease in the 1920s and 1950s caused crop losses of 10 to 20 percent [55].

One of the bigger threats to rice is also a fungus with the common name of **Rice Blast Disease**. This can be found in all regions of the world and can be attributed to a 10 to 30 percent crop loss across the planet [56].

Continuing mutations of these fungi and others that impact the dozen major monoculture crops will cause food shortages, price hikes, and increasing the amount of malnutrition in the world, particularly in poorer developing countries.

The other risk here is that in the fight against these diseases, more pesticides and herbicides will be needed, with the potential of further polluting our soil, waterways, and diet.

The same issue of lack of diversity in our crops applies to farm animals as well. **African Swine Fever** has recently taken hold in Asia and Eastern Europe, infecting over 360,000 animals in 2018 alone. It is highly contagious and there is no vaccine or known cure for this disease. Once detected, the entire herd must be put down (57). Other examples where a sudden outbreak can cause big disruptions to our food supply, due to the limited kinds of food we eat, include **Bird Flu and Mad Cow Disease (BSE).**

Ocean Fish Populations

Without significant intervention and enforced regulation by governments around the world, the current trends of overfishing, pollution, and a changing environment will cause **all** exploitable fisheries in the Asia-Pacific region to collapse within the next three decades (58). The loss of this protein source will not only impact our food supply but the entire ocean food chain.

Already, there are large regions of the South China Sea that cannot support industrial fishing and Chinese fishing fleets are now having to go much greater distances, such as waters near Africa and South America to get their catch.

Another example is the collapse of the Atlantic northwest cod fisheries, where overfishing caused nearly a complete disappearance of cod from the region. After decades of a cod fishing moratorium, stocks have returned but appear to be quite fragile now due to warming ocean currents.

Other Impacts

Here are a few more examples of current impacts that can be directly related to biodiversity loss.

- **Tourism** industries that are related to species in decline such as corals are starting to see a decline in business.
- The **loss of wetlands** is resulting in more damage from overland flooding.

- Pesticides that are used on crops are impacting the insects we rely on for pollination.
- **Forests** that are harvested or destroyed, no longer prevent mudslides or retain snowpack to provide for the steady release of fresh water.

Social Aspects

The trends discussed so far have all revolved around our natural world and are important when the topic of sustainability is brought up. They are all critical and need to be factored in when considering how we must adapt to safeguard ourselves and our close family. However, they are only one part of the picture.

Unless you are a hermit and live completely isolated from society, your daily life will consist of interacting with others and living in the commerce atmosphere. Over the last century or so, trends have emerged in these areas that will have just as much impact, if not more, than the natural trends. Consider the following questions:

- If a piece of news is brought to your attention, either from the internet, your friends, or even a news station, can you trust that it is real?
- Will you be employable in the upcoming decade or so as robots, artificial intelligence, and the offshoring of work come of age?
- Is our current form of democracy still working for all electorates or has it been taken over by corporations and the elite to fulfill their needs?

These "social trends" can cause just as much ecological anxiety as the "natural trends" and must be considered if you want to really plan for a happy and successful future. I have broken the major social trends into eight categories, and they are briefly discussed in the following sections.

Increasing and Migrating Population

Feasible Planet Index

1

How much can you help?
1 = Low impact
9 = High impact

Conservative estimates indicate that by the year 2050, there will be ***another billion*** people on this planet (United Nations World Population Report). Our current population (October 2019) is 7.7 billion; an extra billion will certainly increase the stress on the planet.

Other people on the planet are important to you and it would not be much fun if you were here by yourself. Community, family, social circles, religions, and country affiliations are needed to make life interesting, give us purpose, and to support us through good and bad times.

Though it is hard to measure and does not happen quickly, there are multiple signs that show these relationships are faltering. Nearly half of American children are born to a mother who is not married. Religious affiliations are slowly declining, especially in Europe. There is an increasing trend to bullying, both in person and through social media. These are just some examples of indicators that show wear on our social fabric.

Up to a few hundred years ago, people who lived in a community, in general, held the same beliefs, values, and followed the same moral codes. With the advancement of technology, increasing population, the migration of people, an increased sense of self-entitlement, and globalization, we are now forced to interact with people who may not hold the same viewpoints as our own.

In today's world, your doctor may be an atheist, your boss an African American, your children's teacher a member of the LGBTQIA community, and your local politician may be Asian. To be successful, we will need to find ways to get along and support one another, regardless of these differences, in our ever-increasing global melting pot.

When we look at environmental and social issues, the migration of populations is inevitable (59). Examples include: the current refugee inflow into Europe from the Middle East, asylum seekers trying to enter the United States from countries like Honduras, and even hurricane evacuees who do not return home, since their home and community have been destroyed.

Food is also a real issue. If you live in an industrialized country, you probably have seen a food guide recommending how many fruits, vegetables, etc. that you should eat. If this is what we are supposed to be eating, why should the people in developing countries not have the same diet? If they did, we would have a serious food shortage already (60).

Another issue we face is the aging baby boomer population in industrialized areas. Baby boomers are defined as people born between 1946 and 1964 and is a significant portion of the population. As this population bubble retires, there will be a smaller percentage of the population that will be working and supporting services that they will use. The chart below shows projections of how the retired population will steadily increase from now until about 2050 (61).

Region	% of population over 65		
	2018	2030	2050
Canada	17	22	25
United States	16	20	21
France	20	24	26
United Kingdom	19	22	24
India	6	8	13
China	11	16	24
Worldwide	9	11	16

This means that when we need more money to support underfunded medical and pension plans, climate change initiatives, and pollution cleaning efforts, there will be fewer people to pay for it.

Another part of the migration trend is rapid urbanization. The world's urban population is expected to double between 2010 and 2050 from the current 2.6 billion to 5.2 billion people [62].

4th Industrial Revolution

Feasible Planet Index

4

How much can you help?

1 = Low impact
9 = High impact

With increasing (smarter) artificial intelligence, automation, and technology, many of the jobs that we once took for granted are disappearing. They are either being taken over by machines and computers, are no longer required or are displaced to other areas of the world where the cost of employment (and the cost of living) is much cheaper.

While certain jobs are vanishing, these same factors will also drive new employment opportunities. This transition will cause turmoil in the job market over the next several decades, and exactly where it will end up, no one knows for sure. What is certain is that there will be a large shift, just like there was during the earlier Industrial Revolutions. The difference this time around is that many of the jobs being replaced (by robots and computers) will be good-paying ones and the outstanding question is if the new employment picture will allow most of us to have the same standard of living that we have today.

The automotive industry is a good example. Automation in the factory has already displaced many human workers. Now, new safety features such as automatic braking, frontend collision avoidance, and other safety features are showing to reduce accidents by around 50% [63], impacting auto bodies, mechanics, and insurance claim adjuster jobs. The introduction of electric vehicles over the next few decades will decrease the number of mechanics and auto assembly positions since the number of parts in an electric car is far less than cars with internal combustion engines. Complete autonomous driving, 3-D printing, and ridesharing are also on the horizon; each threatening to further reduce the labor force.

Now, not all of this will happen overnight, but over the upcoming years, we will need fewer people doing these jobs.

There is the flip side as well, with more jobs created in areas such as those that will take advantage of the new 5G networking protocol or those involved in renewable energy. The point is, there will be a dramatic shift of work.

Here are a few more examples of job areas that will be negatively impacted:

- Military
- Fast food restaurants
- Loan officers, investment advisors, legal aid
- Warehouse workers

The above are just a few examples. Recent studies conclude that somewhere between 1/3 and ½ of American jobs are at risk of change over the next two decades [64] [65]. Combine this with an increasing population will result in even more competition for the remaining jobs, especially in developing countries.

This trend could be the one that impacts you the most yet, like climate change, it is not covered extensively by the news media or governments. Perhaps you even think that it is not that serious. After all, we have already seen significant automation in manufacturing, and the internet has enabled extensive offshoring of work, yet the current unemployment rate is low. The issue with this line of thinking is that the overall unemployment rate is not a very good indicator of the types of jobs that people are now doing.

Over the last few decades, there has been a clear shift towards service-orientated work, contract labor, and more people (working in the Gig Economy) doing one or more part-time jobs [66]. While the unemployment rate is currently low, there has also been a clear trend of a shrinking middle class (more on this in the "Increasing Cost of Living" section later in this book).

The following is a list of the major drivers for job transitions over the next few decades:

- The push away from fossil fuels. This includes electric cars overtaking internal combustion engines, renewable electricity generation, and a decrease in oil and natural gas production and refining.
- Better artificial intelligence and robotics are driving more automation into the services arena where judgment and learning are required. Examples include insurance assessments, stock trading, self-driving cars, the health field, and even strawberry picking [67]!
- Increased internet presence. As adoption of internet applications increase, brick and mortar establishments are on the decline. This includes areas like printed newspapers, retail stores, and even colleges and universities that are being challenged by more cost-effective online learning.
- Emergence of the "Gig Economy".
- Cost reduction will allow automation to be introduced in more positions (for example, fast food restaurant cooks).

The whole notion of being employed is also starting to change. Companies are becoming more open to hiring contract employees for specific temporary assignments. Often referred to as the "Gig Economy", there are forecasts that as much as 40% of the workforce in the United States will work as independent contractors by 2040 [68].

With our high and increasing world population, this revolution will be dramatically different from previous revolutions. The sheer scale and complexity of it will push through several decades of intense change.

News Vulnerability

Feasible Planet Index

3

How much can you help?
1 = Low impact
9 = High impact

Being informed about events that impact you is important, but how do you keep up to date and, more importantly, can you trust what you hear and read?

Take supermarket tabloids as an example. While reading them is entertaining, I think most would agree that they are probably not the best place to be getting information that can be trusted in making serious lifestyle, financial, or political decisions.

Where "Fact-Free" or "Fake News" becomes dangerous is when you feel like you should be able to trust the source. This source may be a national news syndicate, a government official or something that you see referenced many times on social media. Consider the following examples:

- In 2018, the president of the United States, **Donald Trump**, who is covered by all news organizations and is widely reported on, personally **averaged 15 false claims <u>per day</u>** (69).
- In the 1990s, the **Heartland Institute**, a "think tank" that is often referenced on traditional news sites, questioned and denied the health impacts of second-hand smoke. Since then, they have worked hard on publishing climate change denial information including **publishing a book that rejects the human role in climate change and sending it out to 200,000 public school science teachers** (70).
- In two months of 2016, a fake news story which purported that President Obama had banned the pledge of allegiance in public schools **was shared and viewed 21 million times on Facebook**. This story was widely propagated by

bots that inferred the story came from major news sites like ABC [71].

- Examples abound of misleading or manufactured news relating to your health, including articles claiming the CDC warned that flu shots are causing a deadly outbreak [72], the danger of common vaccines [73], and exaggerations of the benefits of many vitamins [74].

Not only do these false news stories influence people, but they also add an air of doubt to real news since it is often hard, at first glance, to tell what is real and what is not.

These misleading stories also tend to propagate faster and are read by more people than well-researched, genuine news. For example, a recent MIT study shows that on Twitter, falsehoods reach a deep retweet and sharing level twenty times faster than truthful news [75].

Though social media sites are attempting to filter some false news, there is also a new threat on the horizon, and it is terrifying. With software that is now available, **videos can be made or altered that show things that did not happen** [76]. For instance, a video can now be made that shows a person saying something that they did not actually say. A famous example is a video showing President Obama doing a profanity-laced public service announcement that never happened [77]. Imagine the immediate impact on a stock price if someone released a fake video of a CEO resigning or the ratings of a politician admitting to fraud.

The reality is that, in general, people are good and kind, they work hard, have families and friends, but pay little or no attention to things happening outside of their immediate sphere. The limited news they pay attention to is in areas that impact them directly or interest them.

This works well for most people and they go about their daily lives **feeling like they are informed enough** about critical events, are updated on topics they are interested in, and entertained with the goings-on of the famous elite. They then take this **fractional**

knowledge and make decisions on it. They vote on it, tell their friends about it, and act on it.

Fake news is usually generated for political or corporate gains and is targeted to take advantage of people and encourage them to make poor consumer and political choices.

The rise in distrust of the news, along with the trend of people getting more of their news from the internet, is having another side effect. Local journalists are becoming extinct. There are now vast deserts across the country where smaller communities no longer have local newspapers to watch over and report on local happenings.

When local journalists and newspapers disappear, communities lose out, since local reporting is the primary watchdog for the public offices in the area.

With the underlying forces of decreasing print advertising, increasing free internet content, and declining readership, this trend shows no sign of slowing down any time soon.

The Rise of Entitlement, The Decline of Trust

Feasible Planet Index

7

How much can you help?
1 = Low impact
9 = High impact

Wii-FM (*What's In It For Me*) is a fun acronym for a not-so-fun topic – **Entitlement**. We all have some level of entitlement, but studies and statistics have shown that our Wii-FM signals are getting stronger. This combined with the **declining level of trust** are, I believe, the **primary reasons** for many of the issues we have today in trying to reach an equilibrium with this planet, especially in industrialized nations.

It is not hard to find examples of Wii-FM, I see them in myself and those around me. The big houses, new cars, not knowing or caring about neighbors, divorce, spiraling debt, and underfunded retirement and medical programs; all show an attitude that we feel that **our needs** are becoming more important than the needs of others.

We are becoming less grateful for what we have.

We feel like we deserve more.

We feel like we are entitled to things without having to put in the effort to earn them.

The increasing Wii-FM signal is not just driving over-consumption and reckless spending, it can also be attributed to people feeling an unhealthy level of self-esteem, self-worth, and other narcissistic tendencies. These traits lead people to trust less since they are skeptical of any viewpoint, and the source of it, that is contrary to their own.

Increased levels of Wii-FM are not the only reason for our declining trust levels. Fake news and politicians who do anything for votes are also fueling our declining trust in each other, our communities, and our governments.

The table below shows a few examples of how both trends are developing:

Trend	Evidence
Wii-FM	Average home size has increased from about 1000 square feet in 1950 to around 2500 square feet in 2014 (78).
Wii-FM	The percentage of credit card debt versus income has doubled since 1980 (79).
Wii-FM	The surge of popularity of posting "selfies", use of beauty products, and plastic surgery.
Distrust	Trust in U.S. government fell from 73% in 1958 to 24% in 2014 (80).
Distrust	Trust in each other has declined from 48% in 1972 to 32% in 2016 (80).
Distrust	Trust in mass media has fallen from about 50% in 2000 to 32% in 2016 (81).

Combined, these two trends reinforce the already-natural tendency for people to place a higher value on short-term benefits over long-term benefits. The lack of trust in projections of where this planet is heading, combined with higher than normal levels of self-esteem and importance, are leading people to view any disruption to their way of thinking or changes in their lives as a threat. A few examples of this include:

- The significant percentage of people that are opposed to the pricing of carbon or deny that climate change is a threat or caused by our activities. This includes two recent provincial elections in Canada where Conservative parties, with platforms of abolishing regional carbon taxes, won.
- The recent election in the United States, where Donald Trump's promise to "bring back coal" swung enough voters in key states like Pennsylvania to get him elected.
- Racial and ideological tensions of special interest groups which has led to public intolerance of LGBTQA rights,

various religious beliefs, different ethnic groups, and certain types of immigration.

The danger here is that for the sake of long-term sustainability, things need to change in the short-term, which is exactly the opposite of what Wii-FM and distrust are enabling.

> Golden Rule:
> Treat others as we wish to be treated

The Golden Rule is part of all major religions, and I am pretty sure is also a value that most agnostics or atheists believe in. But are we really applying this golden rule all the time or just when it is convenient for us? What if someone borrowed a bunch of money and then expected **you** to pay it back? What if they put a hundred pounds of carbon in your air or polluted your water and then expected **you** to deal with it?

This is exactly what is happening right now. In part because of Wii-FM and emphasis on short-term benefits, we are borrowing, consuming, and polluting away at a non-sustainable rate and are expecting our kids, grandchildren, and people in countries far away to deal with it.

Your Online Neighborhood

Feasible Planet Index

7

How much can you help?

1 = Low impact

9 = High impact

A few decades ago, if you and your family were in your home or a public place, you rarely had to worry about safety. Today, this is no longer true. You or a member of your family may be in the living room or bedroom and still be in danger. This is because many of us have our own self-made online neighborhood which may include chat rooms, social media sites, blogs, vlogs, texting, news sites, and a wide variety of other www things. Some of these areas are great for learning, used as a creative outlet, entertainment, and networking, but they can also have a dark side that is complete with bullies, propaganda, unvetted news, and lures designed to take advantage of our weaknesses.

Our youth are at particular risk in these dangerous areas and their parents may not even realize they are in these neighborhoods. Unlike our physical world that everyone can see, parts of our online neighborhood can remain hidden. For instance, a recent study reports that about 70% of teenagers admit to hiding some online activity from their parents; this is up from about 40% from 2010 [(82)].

While some of the time we spend in the online world is beneficial, there are several downsides too. Less in-person interaction is one of them. I am certain that you have been in restaurants and have seen couples sitting across the table from one another and both with their heads down looking at their cell phones. There is also increased stress in that you must return texts and "like" social media posts quickly for fear that if you do not, the other person will think you are ignoring them.

The concept of privacy is changing as well at a few different levels. Getting alone time, measured in several hour blocks, away from your online devices is becoming more difficult. There are also numerous programs keeping track of everything you look at, type in, and where you go, and **nothing is forgotten**. This information is marketed to companies so that they can send very targeted advertising directly to you. Employers can use it to see what you are like out of the office and some of it is even used by government agencies to monitor what you are up to, like what was shown in the information leaked by Eric Snowden about National Security Agency surveillance activities.

How much time do we spend in our online neighborhoods? One recent survey found that in 2015, the average grade 12 student in the US spent about 2 hours texting, 2 hours on social media, and 2 hours in general browsing, totaling **6 hours per day** [(83)]. The average adult is only slightly behind this and spends a daily average of about 4 hours on these combined activities [(84)].

Some previously discussed trends, such as news vulnerability, the rise of entitlement, and the decline of trust are sped up because of our expanding online world and the increased time we spend in it. With the advancements in software like blockchain, technology such as eyeglasses with built-in computers, and increased bandwidth due to 5G, this trend **will define** much of our future.

Economy and Government

Increasing Cost of Living

Feasible Planet Index

6

How much can you help?

1 = Low impact
9 = High impact

We all expect and accept price inflation. Increasing prices are reflected in our cost of living and are hopefully matched by increases in our pay so that our standard of living remains roughly the same, or, in the best scenario, gets better.

While the rate of inflation is driven by many things, the cost of dealing with sustainability issues has only just started to become a factor in the last little while. As the environmental trends unfold, they will become a significant driver in increasing the cost of living.

There are direct financial costs to you in dealing with, avoiding, or stabilizing the issues. The following are examples of major drivers that factor into cost of living increases due to unsustainable trends. (Ways that you can lower the financial impact on you are discussed in section two.)

- Increased property and vehicle insurance premiums as insurance companies try to recoup money from large payouts due to flooding, hurricanes, extreme weather, and forest fires.
- Increased health care costs because of more climate- and pollution-related illnesses.
- Cost of carbon-based fuels will increase as governments implement escalating carbon pricing plans.
- Chances are pretty good that you will spend more for clean air and water.
- Increased prices for food due to impacts of climate change, pollution, biodiversity loss, and freshwater shortage.

- A growing percentage of people will be retiring in industrialized countries over the next few decades. This will increase the tax burden on those still working to support this aging demographic, to pay for sustainability stabilization projects, and to repair for damage done
- Tightening regulations to control all kinds of pollution, emissions reduction, and energy efficiency will have costs associated with them and this will be passed on to you, the consumer.

Related to the increased cost of living is the trend of "The rich get richer and the poor get poorer" which is unfortunately true and will most likely only get worse over time. **Currently, the top 1% own just over 50% of the world's wealth** (85).

There are many other data points that show this trend, including stagnant wage growth, despite a strong job market, and evidence over the last several decades of a shrinking middle class while the rich and lower classes are expanding (86). The graph below sums it up well.

Percent change in real income since 1948

The main drivers for the trend seen above are that the rich invest in investments but pretty much everyone else spends most or all their income on day-to-day living expenses. Decreasing opportunities for good-paying employment is also working against the bottom 90% of the population.

How does this impact you? In addition to the obvious lowering of your standard of living, like it or not, many of the actions that can shield you from trends like pollution and climate change cost money. As your disposable income decreases, you will be less able to afford to protect yourself. A small example is good quality air filters for your furnace being much more expensive than lower quality ones. Other examples include having the ability to afford to filter your water for contaminants or take classes to better your chances for good employment.

Currency, Debt, and Interest Rates

Feasible Planet Index

2

How much can you help?
1 = Low impact
9 = High impact

There are many reasons for it, but the current deficit and debt of several of the largest economic countries, including the United States and China, has reached a critical stage.

China is reporting an official debt to GDP ratio of only about 50%. Unofficial estimates such as one recently done by the Institute of International Finance are putting the number closer to 300%[87]!

As of this writing, January 2019, **the current federal debt of the United States is around 22 trillion dollars (79% of GDP[88])** and the deficit for 2019 alone is forecasted to be 980 billion dollars [89]. Interest paid on this debt for 2018 was 371 billion dollars, an amount, as a percentage of GDP, that has not been this high since the great recession[90].

These numbers do not include the debt accumulated at the state level nor underfunded medical, retirement, and pension plans or hidden debt such as the repair cost of crumbling national infrastructure.

The International Money Foundation (IMF) even issued a recent forecast that by 2025, the US debt profile could be bleaker than that of Italy; one of the weakest countries in the European Union[91]. This is in part due to the **United States being the only advanced economy in the world to forecast an increase in debt to GDP ratio over the next 5 years**[92].

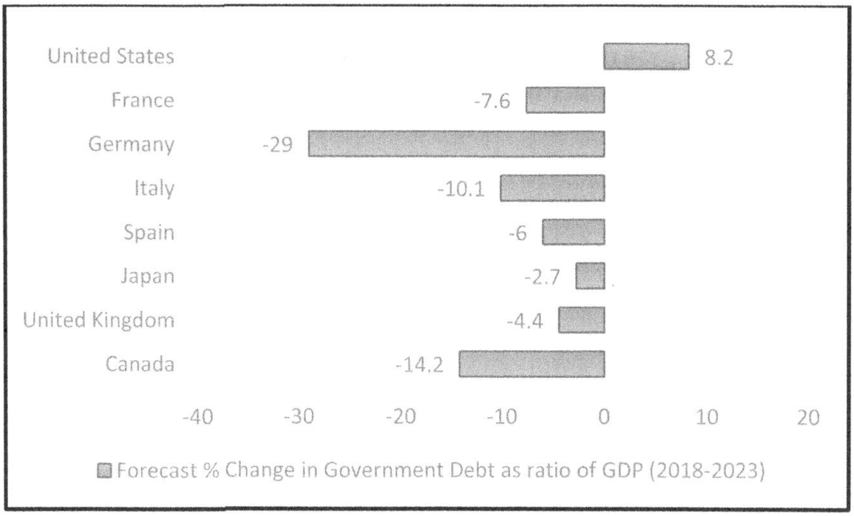

Surprisingly, all of this is happening while the US economy is performing well, with a healthy GDP, no current financial crisis, and **very low interest rates**. This is kind of scary since if any of these conditions change and specifically, if the federal interest rates become higher than the GDP plus about 2 percent, then this debt level will quickly become a burden and may even become unserviceable (93).

One reason that the United States has been able to handle large debts is that it has been the reserve currency for the world since 1944. This means that countries around the world buy and use the US dollar for international trade and investments. This privileged position is changing though with the gaining acceptance of a few other currencies, such as the Euro and Chinese Renminbi as reserve currencies. This trend can be seen with the US dollar comprising 71% of the world reserve currency usage in 2000 dropping to 62% in 2017 (94).

The US dollar, like all other currencies, is a fiat currency, meaning that it is not backed up by any physical commodity. It has value only because people have confidence in it. Everything in the global market is relative, so it is conceivable that a shift in confidence away from the US dollar could decrease its value relative to other reserve currencies such as the Euro.

Another area of concern has been mentioned already, low interest rates. So low that roughly **one third of all government bonds, worldwide, have a negative yield** and it appears a great deal more is on the way. One of the bigger reasons for the negative yield is central banks reacting to slowing economies. The scale of this is unprecedented in modern history and how this will unwind is not at all known, but a sudden and sharp recession is one likely outcome.

So, how could all this impact you and I over the upcoming few decades? Again, no one knows how things will play out, but if the assumption is made that the US debt continues to rise along with the deficit and that central banks will continue with easy money policies, here are some likely scenarios:

- If there is a crisis of some type, the government will be less able to provide funds for a bailout or to stimulate the economy.
- Any effort to reduce the pace of increasing debt will lead to budget cuts in government programs, less health care, fewer subsidies for education, and potentially higher income and sales tax.
- As the government issues more bonds to cover the increasing deficit, there will be an upward pressure on interest rates. Right now, the interest rates are low, but as more bonds are issued, this upward pressure on interest rates will increase. This, in turn, will be a negative driver on the stock market.
- The negative yield phenomena that we are experiencing right now is not sustainable and when it does start to unravel, the bond and stock market will be heavily impacted. This will impact your savings and may result in increasing interest rates which will directly impact any loans you may have, including your mortgage and automobile.
- With a reduced dominance as the world reserve currency, there is a possibility that there will be less confidence in the dollar and hence, higher prices for the many things that are imported.

Corporate Personhood

Feasible Planet Index

2

How much can you help?

1 = Low impact
9 = High impact

In the third installment of my eco-fiction Percipience book series, *2232*, I lightly touched on what I feel is one of the root causes for our poor behavior towards our planet and each other: ***entitlement***, which is the belief that one is inherently deserving of privileges or special treatment. However, did you know that there is another species on this planet where this is not just a belief but where these rights are put in place by law? I am of course referring to the for-profit corporations, specifically the national and multinational variety.

> ***"I see in the near future a crisis approaching that unnerves me and causes me to tremble for the safety of my country. ... Corporations have been enthroned and an era of corruption in high places will follow, and the money power of the country will endeavor to prolong its reign by working upon the prejudices of the people until all wealth is aggregated in a few hands and the Republic is destroyed."***
>
> **- Abraham Lincoln**

You may wonder why I am referring to corporations as living creatures. The reason is that legally, they have many of the same rights as humans do. They can own property, impose their religious viewpoints, fund political parties, and help shape laws and regulations. **This makes these entities, and the few humans that control them, very powerful.**

There are key differences between us and this species as well. Corporations live in the commerce atmosphere versus the breathable one we do. If managed properly, corporations can live forever. Though they may have a physical presence, such as headquarters, storefronts, or factories, they can move fluidly around the globe and there is no physical thing other than a piece of paper, which defines them. Corporations have another thing going for them too, a little term called *limited liability,* which allows, to a great extent, the powerful people behind corporations and the shareholders not to be held directly responsible for the actions of the corporation.

Corporations, in one form or another, have been around for hundreds of years, but it has only been about 150 years since the limited liability, privately held corporations have been enabled by law. The general thinking at the time was that these entities would be an effective means of raising capital to build the infrastructure that would provide goods, services, and employment. The reason for this is that for-profit corporations have, as their primary goal, the need to provide their shareholders with the best return possible on their investment. This leads to a ruthless drive to improve efficiencies, reduce costs, expand product lines, and grow market shares to maximize both near-term profits and long-term growth.

Cost reduction and growth in a commerce atmosphere mean maximizing profits and expansion. This drives the minimal effort required by law for pollution control, the creation of mind-numbing advertising to convince us to buy, and the minimization of compensation to workers. Some multinational companies are now even suing countries if contracts are changed or canceled due to changing environmental regulations and let us not forget that others simply break the law in the name of profits. Volkswagen "fudging" emission ratings is a good recent example, not to mention the trillions of tax dollars that are not collected due to the hiding of money in offshore accounts of many large corporations.

Limited liability reduces the risk for both shareholders and the people who run the corporations. Shareholders can, at most, lose their initial investment, and the board members, for the most part, cannot be held personally liable for debts or litigation that the corporation incurs. There are thousands of examples where the taxpayer is left on the hook for cleaning up environmental disasters that corporations have left behind. In the US alone, there are over 1,300 sites that have been identified for cleanup through a program called *Superfund*, which is vastly underfunded.

If we are to believe everything that we have been told since birth about the democratic system, it is that it is a fair way to run a country: a government elected by the people and responsible for the people. While it may look good on paper, it is also the perfect vehicle for enabling growth for corporations. This is because a vote from a corporate person is more important than a regular human vote. Corporations can contribute large sums of money to the people they want elected. They can afford high-price and effective lobby groups to push for regulations and laws, or the delay of them, to best suit their needs.

Deep down, many of you must realize that there is a real problem here. Created to help us, the for-profit, limited liability national and multinational corporations now control us and sacrifice our well-being for their own growth. Somehow, we have come to believe that corporations need to have the same or perhaps even more rights than we do in order for them to provide the jobs, goods, and services that they supply, for the inventions they come up with, for the capital they can provide for large projects, and for the efficiencies they drive. Somewhere along the way, we have granted them personhood and have allowed them to take over and rule us through the governments they control.

"The corporations that profit from permanent war need us to be afraid. Fear stops us from objecting to government spending on a bloated military. Fear means we will not ask unpleasant questions of those in power. Fear permits the government to operate in secret. Fear means we are willing to give up our rights and liberties for promises of security. The imposition of fear ensures that the corporations that wrecked the country cannot be challenged. Fear keeps us penned in like livestock."

– Chris Hedges, "The Death of the Liberal Class"

Unworkable Political Systems

Feasible Planet Index

3

How much can you help?

1 = Low impact
9 = High impact

I think that nearly all governments seem to have had issues keeping balanced budgets, deciding on controversial topics, avoiding corruption, having resolve to act through on long-term plans instead of short-term quick wins, and keeping the trust of the public. However, these problems seem to be getting worse over the last few decades and there are a few very good reasons for it.

An important change that has been occurring over the last several decades is the rise of influence of corporations and the elite 1% of the population. Through rising political contributions, lobbying efforts, covert direct advertising, and lucrative corporate appointments to retiring politicians and bureaucrats, there is a considerable influence over who we vote for and the actions that our elected representatives take. Usually for the benefit of corporations and the elite, and not in the best interest of the general population.

People are not blind to this and it shows in our decreasing level of trust in governments. During the Eisenhower administration, people who trusted the government, always or most of the time, was around 80%. There has been a steady decline ever since, with that number standing around 18% under the Trump presidency [95].

This distrust, combined with the rise of entitlement in the voting population, is the primary reason why our governments are less effective. People are starting to see a less optimistic future for themselves and their children, they feel taken advantage of by the rich, elite, and corporations. With a "me first" attitude, it is nearly impossible for a governing body to put in long-term plans that do not have an immediate positive benefit. The results, such as ballooning debt, underfunded pension and medical plans, and lack of climate change initiatives are inevitable.

Populist political parties also prey on this fear of government distrust, promising to right the wrongs, but with little substance, or short-term fixes that do not address long-term issues. These parties tend to be nationalistic as well, since, of course, the perception is that other countries are taking advantage of your country. This leads to a strong anti-immigration push and a tightening of tensions between countries, bringing on the real risk of war.

Support for populist political parties and leaders with the same agenda has grown significantly over the last few decades. Germany, Brazil, Hungry, Poland, and the United States are but a few examples where populism is gaining traction [(96)] [(97)].

The drivers of the rise in this style of government are showing no signs of letting up. Examples from previous issues brought up in this book like the rich getting richer, growing entitlement in the voters, and corporate personhood are all intensifying and fanning the flames. This is occurring when we desperately need long-term plans and cooperation to deal with worldwide issues such as necessary migration due to climate change, pollution, and lack of fresh water. The outcome of this is hard to predict, but it seems that conflict, both internally within countries and against other countries, is inevitable.

Further Study – The Trends

The following list of books and websites can be used to get more detailed information on the trends. They are "kind of" in the order presented in this section, but not exactly, so scan through the entire list if you are looking for a specific topic.

	*The Big Thirst: The Secret Life and Turbulent Future of Water* *– Charles Fishman*
	*Sea Sick: The Global Ocean in Crisis* *– Alanna Mitchel*
	*The Thinking Person's Guide to Climate Change: Second Edition* *- Robert Henson*

	*The Ground Beneath Us: From the Oldest Cities to the Last Wilderness, What Dirt Tells Us About Who We Are* *- Paul Bogard*
	*The Hidden Life of Trees: What They Feel, How They Communicate – Discoveries from a Secret World* *– Peter Wohlleben*
	*The Sixth Extinction: An Unnatural History* *– Elizabeth Kolbert*
	*Bowling Alone: The Collapse and Revival of American Community* *– Robert D Putnam*

	*The 4th Industrial Revolution* *– Klaus Schwab*
	*Rise of the Robots: Technology and the Threat of a Jobless Future* *- Martin Ford*
	*The True Story of Fake News: How Mainstream Media Manipulates Millions* *- Mark Dice*
	*The Narcissism Epidemic: Living in the age of Entitlement* *- Jean M. Twenge and W. Keith Campbell*

	*This Changes Everything: Capitalism vs. the Climate* *- Naomi Klein*
	*Corporations Are Not People: Reclaiming Democracy from Big Money and Global Corporations* *- Jeffrey D. Clements*
"Our Planet" Video Series	*A well-done video series on nature and our planet.* *www.ourplanet.com*

Impacts and Readiness

Groundwork

Paradigm Shift

The last section introduced important trends that will shape our world for the next several decades. This upcoming section will see how these trends may impact you and offer suggestions on actions that may be prudent for your situation. The impacts will be covered in the following order:

- Mental Health
- Physical Health
- Employment and Lifestyle
- Social
- Emergency Readiness

Now, before rolling up our shirt sleeves and getting down to work, I think it will be beneficial to first do a brief review of two stories. One is fictional and the other is from real life, but both have a common theme that will be helpful to understand, and useful in getting the most out of the rest of this book.

The first story is the Dr. Seuss book, *The Lorax*, where the world was nearly destroyed and the Truffula trees were driven to extinction by the Over-ler. This happened despite several clear warning signs that the insane production of "Thneeds" was not sustainable and frequent scoldings from the Lorax.

The second story is about Kodak. For nearly the entire 20th century, Kodak was the dominant company in the photography industry. When digital photos were introduced in the 1990s, Kodak spent billions of dollars on research in the digital area but did not capitalize on it for fear of it cutting into their print photography business. Other companies filled the digital gap, the print photography industry collapsed, and Kodak declared bankruptcy in 2012.

In both stories, failures occurred because the Over-ler and Kodak either ignored, did not believe or did not act on clear signs that their world was changing. **We do not want to end up like the Over-ler or Kodak.**

To avoid the same horrible fate, **we no longer can have the opinion that the issues do not exist, are highly exaggerated or are not impacting us right now**. Instead, we need to accept that the challenges are very real, they are impacting us, and the consequences will only get more severe over time.

With this line of thinking, our perspective changes from **hoping** that the impacts will not happen soon to a **realization that they are happening right now and that we must adapt**. I am not saying that we should give up on efforts to reduce our influence on the environment. These efforts are critical to making sure that consequences are not even worse in the future. But this change in perspective brings up the realization that even with these efforts, **changes will occur**. How severely you are impacted will depend, in part, to how much you and your loved ones adapt to the changing conditions.

Like how the issues cover a wide range of areas, there is a broad range of lifestyle choices where adaptation may be necessary. While the general trends are well-understood, how they will impact you specifically is uncertain, so a good deal of flexibility is needed on your part to cover how changes will impact your unique position.

For you and your loved ones to successfully navigate these changing times, **waiting to be told what to do is <u>not an option</u>**. Because of the conflicting agenda of keeping the commerce atmosphere healthy, governments are slow to recognize and act on many of these trends, which is why we are in this mess to begin with! Relying on them to guide your actions will most likely result in adverse consequences for you. This is why I am recommending that to help yourself the most, **taking pro-active steps <u>now</u>** – both in recognizing trends and taking action – is **critically important**.

There is a great deal of information to cover and at times, it may be difficult to keep motivated. If this occurs, pick the simplest action that you can and do it. Even something as simple as buying a water jug for your fridge (covered in the Physical Health section coming up).

Motivation comes through doing. Most people think it is the other way around, that you need to 'get' motivated to 'do' something. While this may be occasionally true, I think you will find that once you start doing something, your motivation level will soar to do more.

After reading the first section, you may feel the need to go "off-grid", become a "prepper" or run out and build a fallout shelter. While these options may sound appealing to some, for many of us they are not practical or desirable. If you are interested in the topic of "preppers", I have included a few excellent references at the end of this section for further information.

For those of you that fear that we are heading toward a zombie apocalypse, the US government's Center for Disease Control has you covered. Check out this site for more information!

https://www.cdc.gov/cpr/zombie/index.htm

Introducing the Online Neighborhood

Our online neighborhood is listed in the previous section as one of the trends that will play a significant role in our lives. The related impacts and corresponding actions you can take are scattered throughout the rest of this book, or at least they were. After reading through the first draft, I felt that there was one important topic related to this neighborhood that needed to be pulled together and introduced upfront – **the introduction of our youth to the online world**.

Having this presented now, I hope, will give a better context when reading the rest of the book and trying to sort out how impacts and actions apply to our youth. If you are not too involved with younger people, feel free to skip ahead to the Mental Health Impact section.

My definition of the online world is anything with a screen. This includes television, video games, cell phones, tablets, and computers. Though television has been around for a while, the rest has only become mainstream over the last four decades and certain aspects, like social media, have only emerged within the last two decades.

This means that we barely have one generation of youth that has been through the online world introduction process. This gives us only a very small sample of young adults to look at to see how the wide variety of introduction methods have worked and is the reason why there is so little guidance out there.

The advice that exists is mostly around crude "screen time" limits and suggestions for some apps that try to restrict what our youth can view. To me, this is about the same as dropping off your 10-year-old child daily in the middle of New York, Los Angeles or Montreal with a fistful of money and telling them you will pick them up in a few hours.

As we will see through the rest of this book, the online world and the neighborhood that our youth develop within it is complex and directly related to several of the trends brought up in the first section. How well our youth will succeed will depend on what type of neighborhood they build and how they act within it. Since our cognitive abilities are not fully developed until we are in our mid-twenties, it makes sense to gradually introduce our kids to this world and allow them to "grow up" in it instead of just dropping them off downtown and hoping they figure things out.

As our kids are growing up in this online world, we want to make sure that these three goals are always met:

- Physical safety
- Exposure to material and situations that match the ability of the youth.
- Not to restrict the great number of opportunities that exist, which can help to prepare our youth for adulthood.

Just like we kid-proof our home by installing latches on doors or having a gun safe, we can also put controls on the hardware, software, and content that our youth may access. Depending on your level of comfort with technology, you may be able to do some or all the suggestions below, but do not be afraid or ashamed to ask for help!

For instance, I consider myself pretty tech-savvy. I have worked with or around computers for 40 years and even though I do have social media accounts, I fully admit that I do not know all the intricacies of the privacy settings. If you find yourself in this situation, find a young adult who does understand (perhaps an extended family member) and ask for help.

Device/Content	Suggestions
Television and Subscription Services	Nearly all televisions and subscription services have parental controls to match the level of programming to children ages. **Use them!**
Portable devices (cell phones, smartphones, tablets, and laptops)	Make sure that location tracking is not enabled (unless you want to track the device) and that privacy settings are strictly set. If there is a camera on the device, make sure that location tagging is **turned off** for this as well.

	Install software that will allow restrictions on what can be accessed and will monitor and restrict access to social media.
Video Games, Apps, and Online Services	In North America, the ratings are set by the Entertainment Software Rating Board (ESRB) [98]. My advice is to follow their guidance. Even though you may think that your child is a mature 14-year-old, they are most likely not at a mental stage that is ready for a video game with a "Mature" rating. Content is generally suitable for all ages. May contain minimal cartoon, fantasy, mild violence, and infrequent use of mild language. Content is generally suitable for ages 10 and up. May contain more cartoon, fantasy, mild violence, mild language, and minimal suggestive themes. Content is generally suitable for ages 13 and up. May contain violence, suggestive themes, crude humor, minimal blood, simulated gambling, and infrequent use of strong language. Content is generally suitable for ages 17 and up. May contain intense violence, blood and gore, sexual content, and strong language. Content suitable only for adults ages 18 and up. May include prolonged scenes of intense violence, graphic sexual content, and gambling with real currency.

	Not yet assigned a final ESRB rating. In addition to a suggestion on an appropriate age, the ESRB also gives content descriptors such as sexual content or violence, and interactive elements such as exposure to unfiltered user-generated content. Ensure that if your child sets up an online account, that their real age is used to ensure age-appropriate content on the site.
Movies and television content	Depending on the country you are in, there are various agencies that set movie and television content. In general, they follow these guidelines: **G** - General – Suitable for all ages **PG** – Parental Guidance – May contain some material that parents may not like for their young children. **PG13** – Parents strongly cautioned that material is most likely not suitable for children under the age of 13. **R** – Restricted – Under 17 requires an accompanying adult. Contains some adult material. **AO** – Adults Only – No one under 17 admitted. Clearly adult content.
Music	Music may have a "Parental Advisory – Explicit Content" label on it. This label means that the lyrics may contain content that is not suitable for listeners 17 and under. In general, pay attention to who your children are listening to. If you are not familiar with the musician, do an online query to see if their songs are age-appropriate.

The above table has the basics for protection and monitoring. This is the easy part, and many parents will stop here, but really all that this achieves is some level of protection for our youth – it does not guide them on how to behave in the online neighborhood, what to watch out for or how it can be used to benefit them. These skills will come through mentoring and requires active participation from you in the area of rules, time limits, participation, and role modeling.

Rules

In the real world, you would probably not allow your 12-year-old to wander around outside at 2 a.m. or to have a secret hiding space in their bedroom that you could not look in. The same goes for the online world and the following list contains suggestions you may want to consider which mimic rules that exist in the real world.

- No electronic devices in private places and must be used in common areas. This includes activities like sleepovers. At nighttime, all portable devices are left in a common place like the kitchen, perhaps have a common charging station there for convenience.
- Never share pictures without permission. If you do not have one set up already, consider setting up a photo site that the entire family and extended family can use to share photos without fear of them getting posted for non-family members to see.
- Never give out real name, address or any other contact information
- No setting up e-mail or social media accounts without your knowledge. Make yourself a friend on all social media accounts.
- Make sure that you know all passwords. At least until the youth gets to a specific age (perhaps 16?).
- Rules on when texting, social media, and other online activity is allowed and when it is not. For example, no texting while walking to school or crossing the street, and no activity at all during supper time.

- Never to reveal a real name, location, phone number etc. even in online games.
- Ensure that your child **knows** that you are monitoring them.
- Have a set of consequences prepared, as best as possible, prior to any issue arising from breaking the rules. For instance, if you find that the youth has created a second, hidden, social media account, smartphone privileges may be lost for a month.

Time Limits, Monitoring, and Restrictions

What does playing Candy Crush, binge-watching television, and floating through Facebook posts have in common? While all of them can be relaxing and fun, they can also be addictive and when done in excess, can hinder us from getting things done that we really should be doing. Youth are particularly hit by this as their ability to have self-control is still developing.

To help them build up good online habits, it is probably not a bad idea to have a discussion around what would be reasonable daily limits for online or screen time activities. When doing this, it is important to keep in mind that some online activities are positive and want to be encouraged, so a blanket "screen time" will not work. Examples of this would include reading an e-book, watching instructional YouTube videos or television shows, or playing an educational game like Minecraft.

Every situation is different, but the table below can be used as a starting point in setting these daily limits. These times may vary daily depending on other activities, weekends or holidays.

	Children under 5	5-13	14-17
Screen time – Total	1 hour or less	Under 4 hours	Under 4 hours
Screen time – Non-productive / entertainment (television, video games)	1 hour or less	Under 2 hours	Under 2 hours

Social Media	N/A	N/A	Under 1 hour
Texts	N/A	N/A	50

Once these limits have been determined, guidelines should also be placed on when these activities are allowed. Some ideas here include:

- No social media at the dinner table or other family times.
- No screen activity one hour before bedtime.
- No texting after 7 p.m. when first getting a smartphone, and perhaps gradually making this later as the youth gets older.

Okay, now that we have sorted out the amount of time, and when that time can be used, the next topic to take care of is restricting where in the online neighborhood our youth can go. There is software that can be used to restrict access and monitor usage, but before setting this up, it is important to discuss with the youth why there is a need for restrictions.

This is important because if your child understands the reasons for the restrictions, they will be less likely to try to get around the rules and any software-block that you may set up or to circumvent you entirely and use a device that is not in your control, such as one belonging to a friend.

Safety is an obvious point to bring up, both their physical safety and the safety of the device from viruses. The other critical topic is the viewing of pornography and you probably need to discuss this sooner than you wish to. The **average age that youth are exposed to viewing pornography is 11**, meaning that many are under this age when first exposed. I am highly recommending getting the book, *Good Pictures, Bad Pictures,* (details in the "Further Study" chapter at the end of this section) or something similar. This specific book is a great start since you can read it with your child and the topic is covered in a natural and comfortable way.

Some suggestions for software include Net Nanny, Kaspersky Safe Kids, and Norton Family. All of these have apps that can be used on smartphones and can also be used on your home computer systems.

Participation & Role Modeling

We cannot effectively raise our children in the physical world without some level of participation, and children look to us as role models on how to behave. The online and screen worlds are no different. Below are a few starting suggestions, several of which are discussed in greater detail later in this book.

- Do not glue yourself to your smartphone while you are with them. Where possible, observe the same social media-free restrictions that you are imposing on your children.
- Help them to set up their social media and e-mail accounts. Trade messages with them. Make sure that you have accounts and are friends with them on all the social sites that they are on.
- Find a game or challenge them to find a game that everyone can play and enjoy. If you do weekly family game nights, make a video game part of the rotation.
- Show how the online world can be beneficial. Look around and find sites, games, and apps that are in areas of interest for your youth. Try virtual reality goggles and online activities that encourage creation such as videos, blogs, books, and music.
- Encourage online collaboration projects.
- Teach them how to evaluate information found on the internet for quality and truthfulness.
- Monitor their messages for the first while to make sure that the content is appropriate and that they are developing good social media skills (see bullets below).

Social Media Etiquette for Youth

1. Post nothing that you would not want your parents, teachers or future employers to see.
2. Only accept friend requests from people you know.
3. Other than your parents, **do not give out your password to anyone**. This includes siblings and friends.
4. Do not reply to messages that harass you or make you feel uncomfortable, even if you really want to. This is the quickest way to disarm cyberbullies as you will show them that you are not concerned about the comment, that they are not very important, and you are in control.
5. Do not use foul language.
6. Do not gossip. This will lower the level of trust people have in you.
7. Do not give out any personal information such as your real name, address, birthday or even vacation plans.
8. Do not post any pictures without permission.
9. Be your own person and do not measure your life based on what other people post. People often post only happy stories online and do not share the boring or unhappy times. Keep things real in your posts and do not let social pressure try to make you something you are not.
10. Do not post about tragic events in our life. Talk about them instead with close friends and family.

More Information

The last few pages were just an introduction to a very complicated topic. If you would like more information on this subject, the books, *Screenwise* and *The Art of Screen Time,* are good starting points. More details on these books can be found at the end of this section.

There are also some great websites for setting up family online use media plans. A great example of one where you can set up a personalized family media plan is at the Healthy Children.org website. Details are in the "Further Study" chapter at the end of this section.

Mental Health Impact

Ecological Anxiety – Adults (18+)

Though the physical and economic impacts of the trends can be easily seen and measured, what is often overlooked are the mental health impacts [99]. Though not all of us may be affected, there is a significant number, perhaps including you, that are, or will suffer from illnesses like depression or anxiety while we struggle to comprehend and live through these alarming times.

The number of people that are impacted by mental health issues and specifically anxiety issues is on the increase [100] and **this topic could be considered to be one of the unsustainable trends** like those discussed in section one. For example, a recent study has shown that three out of four millennials experience ecological anxiety [101].

> One interesting point about anxiety is that about one-third of a person's anxiety level is genetically determined [238].

Personally, I find it hard to imagine many people not having some level of anxiety regarding at least a few of the trends mentioned in this book. The question is to what degree this anxiety exists and how it is impacting you physically, emotionally, and most importantly, the effect it is having on your ability to **act**. Consider the points below, all of which are signs of ecological anxiety:

- Worrying about the state of our democracy, increases in racism, and decreasing opportunities for good employment.
- Helplessness after learning about millions of people in countries far away that have severe malnutrition and lack of drinking water.
- Sadness as you watch animals die or get harmed by man-made structures or pollution.

- Hopelessness by knowing things will get worse and wondering how your children will be able to cope in the next several decades.
- Feelings of nostalgia by seeing things erode and change around you such as glaciers that are disappearing, weeks or months of smoke-filled air, heatwaves, and drought conditions.
- Anxiety over our health, what food is safe to eat, and how and if you will be able to retire due to financial burdens.
- Our unconscious sadness over environmental losses and degradation.
- Overwhelmed by the amount of negative news and projections for both our planet and society.
- Anger at people who spread false news or who do not seem to care about where we are heading.
- Frustration at the ineffectiveness of governments.

These types of feelings and their resulting issues can occur at nearly any age, from the very young to the elderly, but it is the youth that seems to be the most impacted. Dealing with issues in our life is a **learned behavior** and as such, youth are not as well-equipped to handle the weighty issues that they now find themselves immersed in. Ecological anxiety in youth and children are further covered a bit later in this chapter.

Impacts

On the physical and emotional side, ecological anxiety will impact each of us differently. Some of these impacts include:

- Trouble concentrating or thinking about anything other than where our eco-problems are going.
- Persistent worrying about something that you feel you or your community cannot control. Related to this could be a sense of helplessness, sadness or loss of pleasure in activities that you once enjoyed doing.
- Having a sense of impending danger, panic or doom.

- A sense of paralysis or fear of change. Not being able to act even though you know that you should.
- Increased aggression and hostility directed toward people or organizations that are not acting on the trends or are making them worse.
- Loss of social identity, community togetherness, and a sense of isolation.
- Stress in keeping up with your social media. Loss of self-esteem as you perceive others doing better than you based on the exciting lives they post about. Related to this is social media addiction.
- Tiredness, insomnia or restlessness associated with your worry or fears.
- Weight gain or loss, chronic pain, heart disease, and sexual health problems.

At times, it can be hard to self-diagnose, but if any of the above bullets ring a chord with you or if they start to dominate your thoughts, keep you up at night or result in excessive fear, worry or depression – then you may be impacted by some degree of ecological anxiety.

Coping with Ecological Anxiety

It is important to note that this is a very deep and broad topic and the next few pages will only touch on a few key points. I have included a few excellent references for further reading at the end of this section.

The best course of treatment for ecological anxiety is through counselling by a professional; preferably one that specializes in eco-therapy through solution-based therapy (SBT) or cognitive-behavioral therapy (CBT). These professionals can aid you in more balanced thinking and dealing with fear. There are several avenues that you can approach in finding a therapist. This could be through work, your health plan or within your community.

Another avenue is to seek a support group. Finding a local group that focuses on ecological anxiety can be tough but there are many sites online that may be able to help and there are even apps for your phone. The following are a few examples:

The Good Grief Group http://www.goodgriefgroup.org
This group focuses specifically on the indirect impact of unsustainable trends on our mental health. They have developed a 10-point plan, like the AA plan, to help people deal with their anxiety, depression, and other feelings relating back to the erosion of environmental and social aspects of our planet.
Good Therapy https://www.goodtherapy.org/blog/psychpedia/eco-anxiety
A great site with loads of information and links to other resources.
PsyberGuide www.pysberguide.org
A non-profit site that rates the growing number of mental health apps that are out there.

In addition to the suggestions above, there are several activities **that you can do on your own** that may help to reduce your anxiety levels. Hint: Reading through the suggestions below, thinking that some are things you should do and nodding your head **is not** "doing". You must **act** in order to actually **"do"**.

As you go through the list below and the rest of this book, make note of suggestions that will be easy for you to do and then **actually do them. Consciously doing an action, no matter how small, will motivate you to do more.**

Situation/Context: Basic suggestions that may be useful to all adults regardless of their age or level of ecological anxiety.
Do not overload on online news and social media. Everyone has their own circumstances, but 30-60 minutes per day on social media and half that for catching up on news is a healthy amount. If you are obsessed with trying to keep informed on the trends, consider reading well-reviewed books on the topics that interest you. They will usually give a more complete picture of the topic than fractional and unvetted online information. Realize that while the news may be interesting, it can also serve as a constant reminder of the trouble we are in and hence increase your anxiety levels. **Keep in mind that very little news is actionable.** Consider looking for apps or software that can monitor and control the amount of time you spend in your online neighborhood. Examples for your cell phone are Medium, Offtime, and Flipd; all of which have versions for both android and Apple phones. There are also many good choices for computers, a few of which were brought up in the last chapter.
Be slower on text responses. Do not expect instant responses from your message. Get more "in-person" time by scheduling time with your friends and leaving your cell phone at home or in your car.
Get enough rest. No screen time for at least an hour before bed. Limit your caffeine and food intake before going to sleep and try to keep consistent sleep times. An excellent reference book on this subject is, *Sleep Smarter* (details of which are at the end of this section).
Having a healthy diet may be asking too much but at least try to eat less junk food and make sure you are hydrated with water.

Exercise. This can be a gym workout but can also be as simple as a 30-minute walk outdoors each day. Do not do this right before going to bed though!
Situation/Context: Obsessive worrying about gloomy "what if" scenarios.
It is natural to worry about the future to some degree, but we need to make sure it stays at a level that is healthy. It is easy for our brains to make a mountain out of a molehill or to excessively worry about things that we have no control over. *The Anxiety & Worry Workbook* is a good starting point for thinking about your "thinking". Details on this book are at the end of this section.
Not everything about the trends is negative. For instance, there will be new challenging employment and investment opportunities that you may be able to take advantage of. Look for more information on these throughout the rest of this book.
Situation/Context: Feelings of hopelessness.
This may sound a bit twisted but having a feeling of hopelessness over a few or many of the trends can be a good thing. By giving up hope, we are admitting to ourselves that we can no longer rely on someone else to fix the issues and get us off the hook. Thinking about it this way may increase your motivation to prepare and act.
Allow your community to help. Instead of keeping fears bottled up inside, talk it out with friends and co-workers. Get over the stigma associated with talking about your fears.
Situation/Context: Deep concern that you are responsible for one or more of the trends.
No specific action but realize that there are 8 billion other people on the planet that also contribute to the trends. Your parents and grandparents are not innocent either.
Taking an active role in stopping the trends can help to reduce anxiety levels. Can you drive 10% slower on the expressway, buy a bit less clothing or become more active in your community? My

book, *Feasible Planet – A guide to more sustainable living*, is packed with suggestions, many of them super easy.
Situation/Context: Overwhelmed, leading to being frozen and unable to act or prepare.
Just reading books like this one is an excellent start. There is a great deal of information to process here, while at the same time, dealing with your emotions and beliefs. Pick a few easy-to-do things, **do them**, and then go back through this book (or books referenced) and pick a few more actionable items. Rinse and repeat.

Ecological Anxiety – Youth (9-17)

I believe that youth today have more triggers for ecological anxiety than at any other time in recent history. I also believe that the number of ecological triggers and their intensity seems to be accelerating. When compared to youth growing up as little as a few decades ago, consider the following that youth of today must contend with:

- Internet exposure to things they are not mentally prepared for. Associated with this is the never-ending pressure that results from participation in social media.
- Weakening social structure at the family, spiritual, and community levels.
- Constant exposure to technology, such as cell phones and electronic games, leading to less exposure to nature.
- Our youth are getting exposure to information on some trends like pollution and climate change via their school curriculum. Through this, they see that they will bear the brunt of the impact.
- Some youth who understand the impact of the trends, live with parents who deny that the trends are real or are a problem, thus causing tension in the household.
- Though teens are rebellious in general, this is amplified by anger at adults for not doing more about the problems now. There is also anger and conflict with adults who do not believe or realize the seriousness of the issues.
- With limited control of their emotions, exaggerated fear about the future is quite possible.
- With the ongoing turmoil of jobs due to the 4th Industrial Revolution, making career choices is more difficult now than at any other time in the last few centuries.

With all of this, it is no wonder that at least one-third of our youth experience an anxiety disorder [(102)]. Another study showed that four out of five children between the ages of 10 and 12 have experienced fear, sadness, and anger about environmental problems [(103)].

The topic of anxiety in youth is both serious and expansive. The rest of this section will focus on ecological anxiety; the anxiety that can be brought up by the trends discussed in the first section. However, many of the symptoms and suggestions given will apply to any anxiety regardless of the trigger. There are a few excellent resource suggestions at the end of this section that may be of use when encountering anxiety in youth.

Though they seem to grow up quickly, there are a few reasons why ecological anxiety in youth is different from adults. The size of a 6-year-old's brain is 90% of a full-grown adult, but an extensive amount of how it is wired is done during the adolescent growth period.

Though cognitive skills, such as the ability to think through abstract concepts have started to develop by the age of 8, the prefrontal cerebral cortex is still developing. This is the part of the brain that controls emotions, self-regulation, and the ability to think through the cause and effect of actions. Since it is not fully formed, youth usually have limited control over their impulses which can lead to heightened anxiety.

> The cerebral cortex does extensive development during the teenage years but is not fully developed until you are roughly 25 years old!

In general, I believe, that the youth of today also have less developed coping skills such as dealing with difficult emotions, conflict resolution, and handling stress. This results from early exposure to things they were not ready to handle mentally and parenting styles that tend to protect, instead of guide.

Another reason that ecological anxiety in youth may be different from adults is that they are still learning! During their initial exposure and realization of the trends, they do not have all the knowledge to understand the impacts on them. If they start with learning all the negative sides of a situation without hearing about the positive aspects or potential solutions, the unbalanced understanding may cause more fear than is warranted.

Impacts

Many of the symptoms that appear in adults with heightened anxiety also appear in youth. Symptoms that may be more associated with youth are highlighted in the list below:

- **Trouble concentrating or thinking about anything other than where our eco-problems are going.**
- Persistent worrying about something you feel that you or your community cannot control. Related to this could be a sense of helplessness, sadness or loss of pleasure in activities that you once enjoyed doing.
- Have a sense of impending danger, panic or doom.
- A sense of paralysis or fear of change. Not being able to act even though you know that you should.
- Increased aggression and hostility directed toward people or organizations that are not acting on the trends or are making them worse.
- **Loss of social identity, community togetherness, and a sense of isolation.**
- **Stress in keeping up with your social media. Loss of self-esteem as you perceive others doing better than you based on the exciting lives they post about. Related to this is social media addiction.**
- Tiredness, insomnia or restlessness associated with your worries or fears.
- Weight gain or loss, chronic pain, heart disease, and sexual health problems.
- **Quick and unwarranted mood swings.**
- **Self-harm activities such as cutting.**
- **Alcohol and substance abuse, particularly opioids which are cheap, highly addictive, and offer temporary relief from stress and worry.**

Coping with Ecological Anxiety

Most youth spend a great deal of time in their online neighborhood. The combination of social media, computer time, texting, television, and video games make their online neighborhood truly part of their eco-system. Like all neighborhoods, there are good areas and ones that should be avoided as they can be a great driver for heightened anxiety.

Many of the suggestions outlined in the "Introducing the Online Neighborhood" topic in the Groundwork chapter will lower not just eco-anxiety in youth, but anxiety in general. The rules, limits, participation, and role modeling that adults provide will allow a controlled, age-appropriate, and gradual introduction to this world. With this context, it may not be a bad idea to **re-read that section**.

Beyond their online neighborhood, what else can be done to help youth keep their ecological anxiety at a healthy level through their formative years? Several of the suggestions made in the previous section on adult ecological anxiety are useful. These are listed below, with a focus on youth, along with suggestions specific to this age group.

Situation/Context: Basic suggestions that may be useful to all youth regardless of their level of ecological anxiety.
If your youth is obsessed with trying to keep informed on the trends, consider getting them well-reviewed books on the topics that interest them. They will usually give a more complete picture of the topic than fractional and unvetted online or television newscast information.
Ensure that they get enough rest. No screen time for at least an hour before bed, limit their caffeine and food intake before going to sleep, and try to keep consistent sleep times. An excellent reference book on this subject is, *Sleep Smarter* (details of which are at the end of this section).

If you have ever had a teenage boy in the house, you know that they can eat a great deal of food. While girls do not seem to have the same level of appetite, their eating habits tend not to be the healthiest. Encourage good nutrition in both genders and have them avoid fast foods. Ensure hydration.
Exercise! Youth should exercise at least 30 minutes per day.
Situation/Context: Obsessive worrying about gloomy "what if" scenarios.
It is easy for our youth to make a mountain out of a molehill or to excessively worry about things which we have no control over. *The Anxiety Workbook for Teens* and *The Relaxation and Stress Reduction Workbook for Teens* are good starting points for getting them to think about their "thinking". Details on these books are at the end of this section.
Not everything about the trends is negative. Try to get them to see the positive "what if" side. For instance, there are many creative and informative resources for our youth to take advantage of. Think about collaborative music, video or writing projects.
Situation/Context: Education on the trends.
Constantly feed them information on the trends from good-quality sources to counter unvetted information they may have picked up from other sources.
Situation/Context: Feelings of hopelessness.
Ensure that your youth are involved in positive activities around the trends. Put them in charge of household recycling, helping in developing the family budget or in building your family's emergency readiness kit.
This is a fantastic age to empower them to identify and resist the crazy amount of advertising that is targeted to this age group. Check out the website listed at the end for one way to start them learning.

Allow your community to help. Instead of keeping fears bottled up inside, get them involved in youth groups that offer a positive message and help them get over the stigma that is associated with talking about their fears.
Situation/Context: Deep concern that they are responsible for one or more of the trends.
No specific action but just like for adults, help them realize that there are 8 billion other people on the planet that also contribute to the trends. Point out where you and other family members could also do more.
Taking an active role in stopping the trends can help to reduce anxiety levels. My book, *Feasible Planet – A guide to more sustainable living*, is packed with suggestions, many of them are super easy and there is a specific chapter for youth in it.
Situation/Context: Overwhelmed, leading to being frozen and unable to act or prepare.
Help them to pick a few easy-to-do activities. There are plenty of suggestions in this book and in the reference materials. Once these activities are done, pick a few more.
Situation/Context: Involvement in organizations or in preparing for the impacts of the trends.
Get youth involved in helping the family decide what actions should be taken. For instance, have them pick a few ongoing education classes that you could take together.
Encourage involvement in group and community activities. An example is "WE" (info below). Isolation does not help anxiety.
Like adults, encourage them to get involved in correcting the trends, not just adapting to them. Let them know that their actions make a difference.
Get them away from the screen and encourage exercise and outdoor nature activities. Perhaps consider doing activities as a family like raking leaves or having an outdoor picnic.

We are not quite finished with this age group. An upcoming chapter in this book, "**Preparing Our Youth**", covers more aspects that will help our youth be prepared for adulthood.

Interesting Websites, Apps, and Groups for Youth

	13+	Mindshift App uses scientifically proven strategies based on CBT to help you learn to relax and take active steps to take charge of your anxiety. It is available for iPhone and Android and is, as of this writing, free!
	13+	PsyberGuide is a non-profit organization with the goal of helping you find good-quality mental health apps. Psyberguide.org
NASA	All ages	Loads of information on sustainability. climatekids.nasa.gov/how-to-help
	12-18 yrs	Online game where you can build your own sustainable city www.enercities.eu
	6-12 yrs	Website to teach about how to be less influenced by advertising. www. admongo.gov

EcoKids.ca	9-14 yrs	Focusing on environmental awareness and conservation. Games, quizzes, and info pages. www.ecokids.ca

The WE Charity Movement

WE is a youth empowerment organization and worldwide development charity. Active in 10,000 schools in Canada, the United States, and the United Kingdom, the group focuses on young people with programs in service learning, active citizenship, and international development programs.

More info at: www.we.org

Ecological Anxiety – Children (under 9)

Do your young children already have ecological anxiety? After all, they have an active imagination and want to feel safe. Upon hearing of, or experiencing an event from one of the trends, such as a forest fire, they can quickly feel vulnerable, powerless, and scared.

Though there has been a great deal of study in anxiety in adults, ecological anxiety in children is not a well-researched topic. Like youth, anxiety in children is hard to diagnose as it can often be mistaken for a learning disability or attention deficit disorder.

The brain goes through explosive growth in our children. Physically, it is about 90% the size of an adult brain by the time they are 6 years old. The development of skills like speech, curiosity, and the processing of emotions take much longer. The chart below gives a rough outline of how these skills progress.

Age Range	Skills
0-1	Copycat behavior and the beginning of cause-and-effect rationalization.
2-3	The start of independent thinking and answer-seeking.
4-6	Imagination takes off, awareness that their actions impact others, the start of the development of empathy.
7-9	As vocabulary expands, so does their humor, realism starts to become a factor in their thought processes along with the realization that they can manipulate people.

The primary trigger for ecological anxiety in children is a traumatic event such as flooding, hurricanes or even extreme heat. These can be experienced in-person or even by just hearing about it or seeing it on television.

> Increased air pollution may impact mental health disorders for kids as well. There is an increased rate of prescriptions for mental illness in children in areas where air pollution is high [239].

Children have many of the same symptoms that adults have, but there are also symptoms unique to kids. Note that these symptoms are not specific to ecological anxiety and may be related to other factors such as:

- Tantrums, agitation.
- Restlessness, poor focus, disruptive sleep.
- Crying, clinginess.
- Bed wetting, thumb sucking.
- Inability to self soothe.

Regarding ecological anxiety, there are a few goals that we can set for ourselves as we care for this young age group:

- Minimize exposure to traumatic events.
- Re-assure them they are safe if an event occurs.
- Basic education of the building blocks of the trends such as the basics of geography, animals, types of pollution, recycling, money, and how corporations and governments work.

The following are a few age-appropriate suggestions for helping to achieve these goals.

Minimizing Exposure to Traumatic Events
Be safe and do not take any unnecessary chances. Be the first to evacuate if danger is near. Have an emergency fund so that you can ride out a crisis as they occur.
Avoid exposing them to too much second-hand trauma, this means reducing or eliminating their exposure to dramatic stories in the news or avoid discussing these topics with other adults while they are around.
Safety Reassurance
Listen to them to see if they have traumatic events on their minds. Acknowledge their emotions and tell them that it is natural to experience anxiety or sadness.
Explain that for things like pollution and climate change, it will take a long time and we have time to try and do something about it, or at least prepare for it.
Show them what you are specifically doing to help with the topic/concern/issue that they are worried about. Encourage them to do something about it too as this empowers them and you!
Get them involved in the community, family, and good social groups to provide emotional support and help them to build coping mechanisms and confidence in their future.
Answer questions at a high level. Getting into too much detail may overwhelm children since they are just developing the skills to understand cause and effect.
If possible, such as with the issue of climate change, encourage them to do some positive actions to help the problem. Cleaning up plastic pollution is one example.
Teaching the Basics of the Trends
Remember that you are a role model. Kids will quickly adopt your thinking about the risks that the trends pose.
Read to them or have them read good empowering books or watch nature videos. Suggestions are at the end of this section.
Exposing them to nature with activities such as hiking, camping, and gardening.
Do not lie or downplay an issue. Explain using language and terms that they can understand. Ask them what they know about it and how they feel.
Starting at about 6 years old, start teaching them about corporations and specifically about how companies try to make

you buy things through advertising. The website below is a great start.
Starting at about 6 or 7 years old, start teaching them about how all things that they hear may not be true and may be “fake news”.
Starting at about 6 or 7 years old, introduce them to the concept of money and get them involved at a very high level in the family budget.

NAT GEO KIDS	6-12 yrs	Great photos and stories of Earth's peoples and wildlife. kids.nationalgeographic.com
NASA	All ages	Loads of information on sustainability. climatekids.nasa.gov/how-to-help
	Less than 8 yrs	Website with environment games and songs for young children by the Montreal Canada company “Little Animations”. www.littleanimation4kids.com/
ADMONGO .gov	6-12 yrs	Website to teach about how to be less influenced by advertising. www. admongo.gov

Stress

Some issues discussed in the first section do not directly contribute to ecological anxiety, but they do have a direct impact on our mental health in the form of stress.

While the terms 'stress' and 'anxiety' are often used interchangeably, they are not the same thing. The difference between stress and anxiety is that **stress is a response to a situational threat that makes you feel nervous or frustrated**. When the situation is over, the stress goes away too. Anxiety, on the other hand, is a feeling of fear, worry or unease and can be a reaction to stress. Anxiety does not need to be driven by a specific situation but is more of a state of mind that can arise for a multitude of reasons.

An example of the difference between stress and anxiety that relates to the trends is with the 4th Industrial Revolution. Anxiety is your fear of losing your job and not being employable whereas stress would be the event of losing your job which will be relieved when a new job is found.

Another way to think about it is like this: If you are under stress, you can take a break and, at least for a moment, decompress and turn the stress off, the same cannot be said for anxiety. Since anxiety is a state of mind, it is always there at some level, that is until you work to remove the fear, worry, and unease that is causing it.

Some levels of stress are good. It can motivate and drive you into action. Too much stress, however, can result in paralysis in action, anxiety, physical harm, and in extreme cases, post-traumatic stress disorder (PTSD).

Extreme cases of stress like PTSD need professional help. PTSD can develop after a very frightful or stressful event or after a prolonged traumatic experience. With respect to the trends, this can result from extreme weather events that wipe out your home or even an unexpected job loss.

While many of the symptoms and treatments for stress are the same as they are for anxiety, there is one treatment that is different – taking direct action against the event that is causing you stress. Recognizing an issue as stress-related, that it is caused by a specific event, and can be relieved by dealing with this event, allows you to nip the problem in the bud before the stress heightens into true anxiety.

Let us look at a few examples. If you are concerned about how water pollution is impacting you, you can end this concern and stress through installing water filters and purifiers. Stressed about the amount of chemicals in your food? Change your eating habits.

The key is to consciously keep stress at the constructive level and address the issue that is causing it. The following chapters deal with ways you can act and adapt to specific situations to reduce stress and anxiety that are related to the trends.

Physical Health Impact

The knowledge that we have gained in medicine and healthcare over the last century has led to healthier and longer life spans. However, these gains are now at risk of being eroded as the trends discussed in the first section invade our lives in many ways. Through what we consume and are exposed to, this invasion is already well underway and will become a full-out assault within the next couple of decades. If you doubt that there are current physical health impacts, consider the following points:

- **Air pollution** is causing **1,000 early deaths per day** in Europe [104].
- A **heatwave** in 2018 was blamed for the **death of over 70 people** in Quebec Canada [105]. In the same year, a 15-day heatwave in Britain was **blamed for 650 deaths** [106].
- Some vegetables and fruits are laden with **pesticide residue**. Studies have shown that women that eat two or more servings of this type of food per day **are 26% less likely to have a successful pregnancy** than those that eat less than one serving [107].
- About **500,000 children in the US**, between the ages of 1 and 5, have **levels of lead in their blood** that are **higher than the CDC limits** which recommend public health action should be taken [108].

With the planet's predicted population increasing by 2 billion in the next few decades, aging infrastructure, and more extreme weather, the impact to our health will only get worse, especially for children, the elderly, pregnant women, and people who already have chronic diseases. The choice here is to ignore the impacts and allow this to affect you and your family **or** to accept that these are real threats and adapt to mitigate them. Your future and that of your loved ones is up to you.

This section will only cover health impacts that are related to the trends and not all things impacting our health. For instance, there is no discussion on the health risks of some chemicals put into clothes through the manufacturing process or any other unhealthy personal behavior that you may have.

The Water We Drink

Health Impact

Increasing population, manufacturing, and consumption patterns over the last century have resulted in a scary group of chemicals and other contaminants being in our water supply, and trends only show that this will get worse over the upcoming decades.

Basics of Public Water Treatment

Before we get too far into water pollution, I feel that it is important to understand how most of our tap water is cleaned. Roughly 85% of us use water that runs through a public water purification process. The other 15% use private wells for drinking water [(109)].

For most of us, water is taken from rivers, lakes or the ground and then is sent through a series of filters and settling ponds to get rid of the visible particulates like leaves, dirt, and fish. Once through these phases, it is time to get rid of bacteria and other organic matter. This is done mainly through the introduction of chlorine (some cities like Las Vegas use a safer solution, ozone).

Chlorine is very reactive and kills bacteria and other microbes in the water before it is sent through the water pipes and to your sink. As the organic contaminants are killed, the chlorine is used up so ideally, if everything is done right, there will only be a small amount of "free" chlorine left in the water by the time you turn on the tap.

The Pollutants

The following table outlines some dangerous water pollutants and their potential impact. Note that there are other water-related risks. The ones below are related to our unsustainable trends discussed in section one.

Pollutant	Potential Impact
Lead	No amount of lead in water is safe. Lead accumulates in the body and is especially harmful to children and has been linked to damage to the central and peripheral nervous system, learning disabilities, shorter stature, impaired hearing, and impaired formation and function of blood cells.
Nitrates	Nitrates from fertilizers and manure that end up in drinking water increase the risk of colon, kidney, ovarian, and bladder cancers.
Trihalomethanes (THM) or Total Trihalomethanes (TTHM)	TTHM is the result of chloride interacting with organic matter. Put simply, the more organic matter, the more TTHMs. At high enough levels, TTHMs are associated with increased risk of cancer along with bladder, liver, kidney, and intestinal tumours. Some studies suggest that TTHMs increase the risk of problems during pregnancy, including miscarriage, cardiovascular defects, neural tube defects, and low birth weight.
Arsenic	Long-term exposure to arsenic in drinking water can cause cancer and skin lesions. It has also been associated with cardiovascular disease and diabetes. In utero and early childhood, exposure has been linked to negative impacts on cognitive development and increased deaths in young adults.
Chlorine	Chlorinated water speeds up the aging process, like the effects of extended exposure to the sun by depleting skin and hair of natural oils. It can promote acne and other skin irritations, including eczema and rashes. Drinking chlorinated tap water has been linked to bladder, breast, and bowel cancer; and long-term effects of drinking, bathing, and swimming in chlorinated water has been shown to cause malignant melanoma, otherwise known as skin cancer [110].
Chromium	In addition to cancer and reproductive harm, short and long-term exposures to chromium can lead to eye and respiratory irritation,

	asthma attacks, nasal ulcers, dermal burns, anemia, acute gastroenteritis, vertigo, gastrointestinal hemorrhage, convulsions, ulcers, and damage or failure of the liver and kidney [111].
Per- and polyfluoroalkyl (PFAS)	PFAS are used in a variety of industrial processes and products throughout the world. There are limited, credible studies to date on PFAS but several indicate that in humans, exposure to PFAS may affect growth, learning, and behavior of infants and older children. In adults, it may lower the chances of getting pregnant, increase cholesterol, increase cancer risk, and affect the immune system. Unlike some pollutants, PFAS do not break down in your body but instead accumulate over time.
Microplastics	The primary concern with microplastics is that they are carriers for unwanted toxins and enter the bloodstream.

Measurements

Federal environmental agencies set out guidelines for the acceptable levels of the various types of pollutants. Water agencies keep our water safe by measuring the level of contaminants and adjusting the purification process as needed to keep within the guidelines. For instance, in the springtime, when there is a lot of runoff and organic material in the water, agencies usually increase the amount of chlorine.

Unlike air pollution, where the key contaminants are widely and frequently monitored, the levels of water contaminants are not as closely watched and the data is harder for the public to view and understand. Unlike air pollution which has an Air Quality Index (AQI), there is no single simple index for water quality.

What you can do

While the water you drink from a public water supply should meet safety regulations, with the complexity of water filtration and distribution, **there is no guarantee**. Pregnant women, children, and the elderly are the most susceptible to pollution in our air and water. However, unlike fast-acting air pollution, water pollution tends to affect our long-term health. A recent example is the lead contamination in Flint Michigan.

Below are several ways that you can reduce your level of exposure to water contaminants. Even if you are not too worried about your health, there are several easy-to-do and cheap options that will reduce the contaminants in your water. We will start exploring these options first and then build up to more elaborate solutions.

Let us start by looking at reducing the chlorine that you are exposed to. Chlorine evaporates very quickly. **Filling a jug with water and keeping it in the fridge for 24 hours will eliminate most of the chlorine in the water**.

Many other contaminants that are found in your water can be reduced by purchasing a water pitcher with a built-in filter. A good quality pitcher will reduce or remove many contaminants such as lead, mercury, copper, fluoride, chromium, chloramines, microplastics, THMs, and chlorine. Check the specifications to ensure that it will filter what you want it to. Note that, as of this writing, I cannot find any pitchers that are certified to remove PFAS.

> A quick note. Boiling water will help kill organics in your water and in removing chlorine but some chemicals, like fluoride, are not removed. What you are left with after boiling is water that has a higher concentration of these types of contaminants since some regular water has evaporated through steam.

Another option is to use purchased bottled water, either bought in a store or delivered to your home. Though most bottled water has gone through rigorous purification, there are some brands that simply bottle publicly cleaned water so will still retain chlorine and other chemicals. Also, most (perhaps as high as 90%) of bottled water contains microplastics.

If you do decide to get a water jug with a filter or a home water filter system, also consider getting a reusable water bottle for when you are on the go and fill it with your clean water. There are some very nice reusable water bottles on the market now. Some that will keep your water ice-cold for 24 hours!

An easy but relatively expensive way to reduce your exposure to many of the pollutants in your drinking water is by buying a water purifier for your home. There are several types on the market:

- Reverse osmosis is effective for removing lead and is probably your best bet at removing, or at least reducing, PFAS exposure.
- Reverse osmosis and ion exchange filters will remove microplastics.

Your skin can absorb chlorine very quickly. A ten-minute shower or a 30-minute bath is equivalent to drinking two liters of chlorinated water [(112)]. If you like hot showers or baths, then the amount of chlorine that you breathe in via steam is even more than your skin absorbs. This is particularly bad since it bypasses the digestive system and goes straight into your lungs. Several quick fixes for this include taking shorter baths and showers, using a bit cooler water, taking showers less frequently, and installing a water filter on your shower head.

For me, the shower head filter is a great solution. They are inexpensive to buy, and the filters usually only need to be replaced once or twice a year. They are easy to install and many not only eliminate most of the chlorine, but they will also remove other harmful contaminants as well while softening your water a bit.

If your water is from a private well, then you are responsible for keeping your water safe. Test kits can be purchased to check your water or samples can be sent to laboratories for analysis.

Even if you have publicly supplied water, you may want to test for certain chemical levels. Water hardness is a common one and levels of chlorine can be easily tested with a common swimming pool test kit.

If you have an older house, check to see if it has any lead water pipes and have them replaced. This may not completely free you of lead exposure though as older city water mains can contain lead as well.

Chlorine works great for keeping your swimming pool water safe but can be hard on the eyes and skin. As an alternative to chlorine, and especially if you have young children using your pool, you may want to consider alternatives such as bromine.

The Air We Breathe

Impact

Despite Mother Nature's best efforts at trying to keep it clean, our air quality is just getting worse. The primary health worries for us are increasing ground-level ozone levels, extra particulate matter, higher aeroallergen levels, and additional dust.

How are these impacting you? Though everyone is impacted to some degree, those affected most are children, the elderly, pregnant women, and people with chronic diseases such as asthma, chronic lung disease, heart disease, and diabetes. Impacts vary greatly between people and your level of activity also factors in. For instance, an outdoor laborer or jogger will be impacted higher than an office worker.

Even short-term exposure to air pollution for pregnant women can increase the risk of premature birth and lifelong respiratory and neurological in their children [(113)]. Even if your local air is relatively clean, it is important to understand the types of air pollution, impacts, and ways to protect yourself in case you travel to more polluted areas.

Long-term exposure can cause increased blood pressure [(114)], accelerated aging of your lungs, increased probability of getting asthma, bronchitis, cancer, and a shortened life span [(115)].

Pollutant	Description	Potential Impact
O3	Ozone	Reduced airflow to lungs, can cause coughing, throat irritation, and difficulty in breathing. In severe cases, it can cause hospitalization or death.
H2S	Hydrogen Sulfide	Low concentrations can cause headaches, memory loss, tiredness, and balance problems. At higher

		concentrations, loss of consciousness or death.
NO2	Nitrogen Dioxide	Increases likelihood of respiratory problems as it reduces immunity to lung infections. Coughing, wheezing, colds, flu, and bronchitis.
SO2	Sulfur Dioxide	Irritates the nose and throat, shortness of breath, and tightness in the chest. Higher concentration can result in death.
PM10	10-micron or less particles like dust and ash (PM stands for particulate matter)	At lower concentrations, coughing, wheezing and at higher concentrations, high blood pressure, heart attacks, and death. **Any** concentration will hurt your health.
PM2.5	2.5-micron or less	Can result in breathing problems including asthma and heart issues. Promotes cancer.
VOC	Volatile Organic Compounds (examples include benzene, formaldehyde, and ethylene).	Eye, nose or throat irritation. At high concentrations can cause headaches, loss of concentration, nausea, and damage to kidneys, liver or central nervous system.

Measurements

The most common gauge of air quality is represented by the Air Quality Index (AQI) (see bottom of this section for exceptions, such as in Canada). The AQI index is a single number between 0 (good) and 500 (bad) that represents the overall risk of the common air pollution components. It is computed on a real-time basis in thousands of locations around the globe and is most likely available in your detailed weather forecast. An excellent resource for seeing AQI readings around the world is at this link:

https://aqicn.org/map/usa/

The pollutants that are included in the AQI index vary a bit depending on where you are in the world, but at a minimum covers ground-level ozone, nitrogen dioxide, and particulate matter under 2.5 microns ($pm_{2.5}$). Some AQI's also include sulfur dioxide and pm10 (particulates under 10 microns in size).

The following chart outlines value ranges for the AQI index along with risks at the various levels.

AQI Value Range	Risk Level	Caution Statement [116]
0-50	Good	None
51-100	Moderate	Active children and adults, along with people with a respiratory disease should **LIMIT prolonged** outdoor exertion.
101-150	Unhealthy for sensitive groups	
151-200	Unhealthy	Active children and adults, along with people with a respiratory disease should **AVOID prolonged** outdoor exertion. Everyone else should **LIMIT prolonged** outdoor exertion.
201-300	Very Unhealthy	Active children and adults, along with people with a respiratory disease should **AVOID ANY** outdoor exertion. Everyone else should **LIMIT ANY** outdoor exertion.
300+	Hazardous	Everyone should **AVOID ANY** outdoor exertion.

Some literature and regulations have pollutant levels stated in parts per million (ppm) or micrograms per cubic meter. There is a handy calculator at this link that you can use to convert these numbers to AQI Index values.

https://airnow.gov/index.cfm?action=airnow.calculator

If you live in the United States, you can retrieve historical AQI numbers at the following site:

https://www.epa.gov/outdoor-air-quality-data/air-quality-index-report

From this source, here is 2017 data for a few select cities that have the number of days in each AQI Value range.

City	Good	Moderate	Unhealthy for Sensitive Groups	Unhealthy	Very Unhealthy
Boston	178	182	5	0	0
Chicago	157	183	22	2	0
Denver	72	253	39	1	0
Houston	184	156	22	3	0
Los Angeles	38	205	76	38	8
Miami	213	137	8	0	0
New York	154	192	17	2	0
Phoenix	32	238	82	11	2
Seattle	222	119	13	10	1

Not all regions report air quality with AQI but use the Air Quality Health Index (AQHI) instead. Examples are Canada and Hong Kong. The AQHI is like the AQI but it uses a different numeric range. The table below shows the AQHI value ranges and their meaning [117].

AQHI Value Range	Risk Level	Caution Statement
1-3	Low	None
4-6	Moderate	Active children and adults, along with people with a respiratory

		disease should **LIMIT prolonged** outdoor exertion.
7-10	High	Active children and adults, along with people with a respiratory disease should **AVOID prolonged** outdoor exertion. Everyone else should **LIMIT prolonged** outdoor exertion.
Above 10	Very High	Everyone should **AVOID ANY** outdoor exertion.

The chart below shows AQHI values for a few select Canadian cities (number of days in AQHI range).

City	Good	Moderate	Unhealthy for Sensitive Groups	Unhealthy	Very Unhealthy
Calgary	347	18	0	0	0
Toronto	297	67	1	0	0

Air masks

You see people wearing them but do these masks help? The answer depends on the types of pollutants you are trying to avoid and the type of mask. For example, a normal N95 mask will be effective against $PM_{2.5}$ but is useless against filtering ozone.

The air masks are classified with both a letter and number rating as shown in the tables below.

Air Mask Rating Letter	
N	Not oil resistant
R	Resistant to oil
P	Oilproof

Air Mask Rating Number	
95	Peak efficiency is 95% blockage of 0.3 microns or larger.

99	Peak efficiency is 99% blockage of 0.3 microns or larger.
100	Peak efficiency is 99.97% blockage of 0.3 microns or larger.

Note that the above numbers are under perfect conditions. If your mask does not fit well, the effectiveness can drop quickly. For example, men with beards or who are unshaven will have nearly an impossible time to find a normal mask that will fit and be effective.

Here is a table of sizes for common airborne particles and a few reference items.

Particle	Size (microns)
Anthrax	1 - 5
Asbestos	0.7 - 90
Bacteria	0.3 - 60
Beach Sand (fine)	90
Coal Dust	1 - 100
Fiberglass Insulation	1 - 1000
Human Hair (average)	50 - 70
Mold Spores	10 - 30
Ozone (and other gas molecules)	Less than 0.001
Pesticides & Herbicides	0.001
$PM_{2.5}$	2.5 or less
PM_{10}	10 or less
Pollen	10 - 1000
Radioactive Fallout	1 - 10
Tobacco Smoke	0.01 – 1
Viruses	0.004 – 0.3
Wildfire Smoke	0.4 – 0.7

The answer to the question about if you should wear an air mask and if so, what kind, depends on many factors. In general, if the AQI/AQHI is above normal, is driven mostly by particulate matter, and you will be spending a fair bit of time outdoors, then an N95-type mask is probably a good idea. There is little reason to go higher than an N95-type mask for normal casual daily activities. The higher-rated masks are more for work-related activities.

If your high AQI/AQHI reading is mostly because of ozone, nitrogen dioxide or sulfur dioxide, then a normal air mask **will not** do you much good since the gas molecules are much smaller than the filter size of the mask. If you want to reduce exposure to VOCs and ozone, you should look at getting a mask with **carbon filtration** (more on carbon filtration of ozone and VOCs a bit later in this section).

Quality of masks vary but nearly all can be used multiple times, depending on the level of pollutants and activity. It is critical to find one that will fit your face well since the many gaps between the skin and mask will cause the efficiency to drop quickly.

Besides reducing the amount of air pollution, air masks can also help to prevent the spread of viruses and exposure to mold and pollen spores. Spores are usually larger than the size of the filter and viruses are attached to mucus, again, much bigger than the filter size.

When bought in bulk of 10 or more, the typical price of a mask is less than $2 USD. There are small size N95 filters that can be purchased for children as well and tend to cost a bit more.

Air purifiers

If your air quality is normally not that great, you may want to consider some type of home air filtration. For those of you that have a furnace and/or air conditioner, this can be done with an air filter or a standalone unit. If you live where there is central heating, such as an apartment, then your only choice is the standalone units.

Before discussing the options, here are some common terms used in this area:

Term	Definition
MERV (Minimum Efficiency Report Value)	A standard for the size of particulates that a filter will remove (furnace, A/C, vacuum

	cleaner). Home use ranges are from 1 through to about 16. A chart in the next section shows the different filtering capabilities of the ratings. Higher MERV ratings such as 17-20 are for medical and industrial applications.
HEPA or True HEPA (High Efficiency Particle Absorber)	Will capture 99.97% of particulates that are 0.3 microns or larger. Equivalent to MERV 17 rating.
ULPA (Ultra Low Particle Air Filter)	Will capture 99.999% of particulates that are 0.1 microns or larger. Equivalent to MERV 20 rating.

"True HEPA" is a marketing term to show that the filter meets the HEPA standards. Any other HEPA term such as *HEPA-style, HEPA-like* or *99% HEPA* indicates that the product probably does not meet the HEPA standard.

Forced Air Furnace/Air Conditioner Filters

Forced air furnace and air conditioner filters work by forcing air through a dense set of fibers. The tighter the fibers, the finer the particulates that are blocked. A MERV rating between 8 and 11 is adequate for most homes.

* It is important to note that a higher efficiency air filter like a HEPA rated one may reduce the air quality in your home if your furnace is not rated for one since the airflow will be reduced. Ensure that you do not exceed the MERV ratings for your furnace and that you clean or replace your filter regularly.

The following is a table with guidelines for normal household filters and what they filter out.

MERV Rating	Typical items filtered
1	Textile and Carpet Fibres, Sawdust

2	Sanding and Spray Paint Dust
3	Dust Mites, Pollen
4	Particulates greater than 10 microns
5	Cement Dust
6	Fabric Protector
7	Mold Spores
8	Particulates greater than 3 microns
9	Auto Emissions
10	Flour Dust
11	Land and Humidifier Dust
12	Particulates greater than 1 micron
13	Microscopic Allergens
14	Most Tobacco Smoke
15	All Bacteria
16	Particulates greater than 0.3 microns

Alternate Filter Rating Systems

To make sure things are not too easy for consumers, a few companies have ditched the MERV rating and have developed their own rating guide. Home Depot® uses Filter Performance Rating (FPR) and 3M® uses Micro-Particle Performance Rating (MPR). Each of the rating systems measures performance a bit differently, so a perfect comparison between them and MERV is impossible, but the chart below is close [118]. **Values in bold** should provide adequate filtration for the average home.

MERV	MPR	FPR
5	300	
6	300	
7	600	5
8	**600**	**5**
11	**1000**	7
11	**1200**	7
12	1500	10
13	1900	10
13	2200	
14	2800	

Ozone Generating Air Purifiers

Ozone is an unstable oxygen molecule that will react with pretty much anything it runs into. This exceptional reactivity makes ozone highly effective at removing odors, molds, mildews, and airborne chemicals. However, this reactivity also allows it to react with our skin, eyes, and lungs causing, depending on the concentration, damage both in the short and long term.

This health risk has resulted in the CARB (California Air Resources Board) recommending that **ozone generators, which are machines that intentionally produce ozone, not be used in the home** [(119)]. Some household ozone air purifiers are banned for sale in California and there are warnings from the Environment Protection Agency (EPA) as to the risks.

Ozone purifiers that emit levels of ozone that do not exceed public health standards are not very effective at reducing airborne containments [(120)] anyhow, so why take the risk?

Activated Carbon Filtration (also referred to as a charcoal filter)

Carbon filters work by removing impurities and contaminants from the air through chemical adsorption (filter does not swell). There are two types: powdered block and granulated, with the powdered block being most effective. Some interesting properties of activated carbon filters are:

- Is effective on vapor-based contaminants and odors such as chemicals from clothes that have been dry-cleaned, household cleaners, and pesticides.
- Will absorb organic molecules, especially solvents.
- Have a large capacity to catalytically destroy ozone and VOCs.
- Not very effective in filtering particulate matter that is not vapor-based like dust, pollen, spores, and sand.

Carbon filters can be purchased as a standalone unit or as part of another filter such as an air mask or furnace filter.

Do you have a stinky fridge or want to remove room odors, including some VOCs, on the cheap? Place six or so BBQ charcoals in a bowl, pan, old sock or pie container and place this in the fridge or the targeted room. Use natural briquettes, not the easy-light kind. Replace them every 2-4 weeks and remember to use the "used" ones in your BBQ!

Ionic Air Purifiers

Ionic air purifiers (sometimes referred to as ionizers) work by using high voltage to electrify (ionize) air molecules. Airborne particles become charged when they are in contact with the ionized air molecules and then are captured by earthed grounding plates in the purifier or to other surfaces such as walls and ceilings. This is different from ozone purifiers which rely on a chemical reaction.

Ionizers can offer an effective solution for reducing asthma and allergy symptoms but are not very effective for removing VOCs, ozone or odors. According to the EPA, ionizers typically have low effectiveness because of very low airflow rates; they also generate some level of ozone [121].

UV Air Filtration

If you are really concerned about airborne pathogens, then you may want to consider an ultraviolet air filtration. These can be purchased as standalone units or as an additional piece of hardware to your furnace or A/C system.

These systems work by killing, not removing, the bacteria, viruses, and mold that are in your air.

Reducing Home VOC (Volatile Organic Compounds) levels

VOCs can be found in all kinds of products such as household cleaners, indoor pesticides, glue, paint, varnish, cosmetics, and hairspray. As these products evaporate, they release VOCs, compounds such as benzene, trichloroethylene, and formaldehyde.

The best way to reduce VOC levels in your home is to increase the airflow through your home and buy fewer products that emit them. A list of products that commonly contain VOCs are shown below. Since this is a general list, check the label when purchasing a product to see if it specifically contains the VOC listed.

VOC	Products commonly found in
Acetone	Nail polish remover, furniture polish, wallpaper
Benzene	Paint, glue, some carpets
Butanol	Candles, gas stoves, barbecues, cigarettes
Dichlorobenzene	Mothballs, deodorizers
Ethanol	Glass cleaner, detergents (both dishwasher and laundry)
Formaldehyde	Wood lacquers, some wood and plastic products
Terpenes	Fragrant products such as soap and laundry detergent
Toluene	Paint
Xylene	Idling cars

If you want to filter VOCs, the only effective method is by using activated carbon filters. If you have a concern about the level of VOCs in your home, single-use test kits are available for formaldehyde at a reasonable price (less than $20). There is more extensive home testing equipment available as well but will run you at least a few hundred dollars.

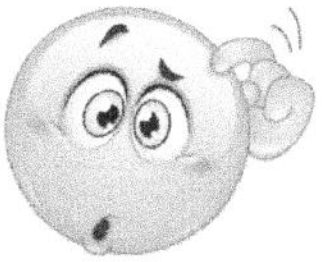

Contrary to what you may think, read in books or what websites may report, indoor plants do not have any meaningful impact on the level of volatile organic compounds (VOCs) in your home [211] [212].

Car cabin air pollution reduction

Clogged freeways in the baking heat during afternoon rush hour may be one of the worst places that many of us experience. Not only do the idling cars and large transportation vehicles spit out harmful emissions, but the heat also turns some of these emissions into harmful ozone gas.

The most cost-effective way to reduce some of this pollution is to ensure that your car's heating and air conditioning system is set to recirculate the air instead of bringing in fresh air. Another way would be to wear an air mask during the heaviest commuting times.

Most cars built after the year 2000 also contain a replaceable cabin air filter. These can usually be replaced by yourself and have the same characteristics as home forced air filters. You can even buy ones that have active carbon filtration combined with the normal particulate removal. Carbon filtration is important if you are exposed to high levels of ozone or VOCs while driving or sitting in rush hour traffic.

App	Description
Air Matters	Free app from the U.S. EPA and gives air quality information.
Plume	Another nice app for detailed weather and air quality data.

EWG's Healthy Living	A nice app that allows you to scan a product and get health-related information (such as VOCs).

Our Food Supply

Our food supply cannot avoid being contaminated by the pollution in our soil, air, and water, however, these are just the start of a list of health factors related to the food you and I consume today. Corporate personhood, biodiversity loss, fake news, climate change, and the shortage of freshwater also contributes to food contamination or in influencing you into making unhealthy choices.

Since everyone is different in their diet preference, what is available to them, and what they can afford, this section does not layout a specific healthy diet plan, rules around sustainable food growth or ethical farming. Instead, the health impacts from the topics outlined in the first section of this book are given along with choices that you can decide to incorporate (or not) into your specific food intake.

> Did you know?
>
> A recent study found microplastics in 90% of the 39 salt brands tested [244].

For example, this part of the book does not get into the pros and cons of eating red meat but does talk about the increasing use of antibiotics and hormones throughout our food supply chain because of the industrial food complex's need for profit and our growing population.

If you are interested in a healthier diet or supporting sustainable agriculture, there are a few excellent book recommendations at the end of this section.

Organic Foods

The definition of 'organic' varies but in general, for plants, it means that the food was grown without certain pesticides and fertilizers, not irradiated, and is grown in a way that supports biodiversity and the ecological balance. Many certifications, like the USDA organic certification, prohibit any genetically modified foods from being labelled as organic.

For meats – the general definition of organic is that no hormones are given to the animals and that their feed contains no animal by-products, antibiotics, GMO grains or grains grown with persistent fertilizers or pesticides.

In general, these rules are good for the environment. Biodiversity and soil management, key points in organic farming, are critical to our future and the reduced use of chemicals is good for our soil, water, and ***sometimes*** us.

The trick with organics is that the industrial food complex has spent a great deal of money and has convinced most of the population that organic food is better in nutritional value when compared to their non-organic counterparts. This perception allows them to charge a substantial markup for the organic label.

While ***some*** organic foods may have less chemical contamination, the **organic certification does not mean they have better nutritional value**. The certification simply means they were grown under certain regulated conditions [(122)].

The extra money that can be made by selling organic food also is a motive for cheating the system, especially for imported food which is difficult to verify for several reasons and there are multiple places for errors to occur [(123)]. This is one good reason to, where possible, buy locally grown organic foods where the chances of "errors and omissions" are less likely to occur.

In some cases, organic may not mean what you think it does. In the United States, hydroponic farms can get USDA organic certification, meaning that those organic cherry tomatoes that you just paid a premium price for may have been grown with just chemicals and artificial light.

One last point to clarify related to organic vegetables. Heavy metals like lead and arsenic can still exist in organic foods such as rice-based baby food. This is because these metals can remain in the soil for many years and there is no organic specific regulation around this. Also, though there are regulations around the water that can be used in growing and cleaning organic food, they basically say that it can be tap water, thus other contaminants like those discussed in the water section can still be in your food (124).

Okay, enough of giving organics a tough ride. Keeping toxic chemicals out of your body is important! There are reasons that you may want to consider paying that higher price and buying organic since non-organic farming methods do use a pile of chemicals in growing and processing many foods.

The table below contains two lists. The one on the left are foods that are typically grown and processed with many chemicals, and you will benefit if you buy certified, non-imported organic (125). The list on the right contains foods that typically have few chemicals used in growing and, therefore, make less sense to buy organic, especially if you are on a tight budget (126) (127) (128).

Food and drinks with the **most chemicals** used in growing and processing (**most benefit from buying organic**).	Food and drinks with the **least chemicals** used in growing and processing (**least benefit from buying organic**).
Coffee	Avocados
Strawberries	Broccoli
Spinach	Cauliflower
Apples	Cantaloupe
Grapes	Onions
Peaches	Asparagus
Cherries	Sweet Peas
Celery	Cabbage
Tomatoes	Quinoa
Potatoes	Mushrooms
Sweet Bell Peppers	Carrots (peel before eating)
Pears	Kiwi, Pineapple, Mango, Papaya

On the meat side of things, what we should be careful of is the increased use of antibiotics and growth hormones. These are being used by the industrial food complex more and more to allow for quicker-growing animals in closer confined spaces.

For **antibiotics**, the risk is not so much in ingesting the antibiotics in the meat but has more to do with consuming bacteria that has grown resistant to antibiotics. This is a serious issue with about 2 million people each year in the U.S. getting infected by bacteria that are resistant to one or more antibiotics. From this 2 million people, 23,000 die [(129)].

The safest way to get around this is to eat only meat that has been cooked to the proper temperature by checking with a meat thermometer.

Eating products such as milk and meat where the animal has been given **growth hormones** is a bit more of a serious situation. These hormones can be transferred to your body which may increase your risk for cancer or result in you being resistant to certain types of drugs.

The only effective way to reduce the intake of growth hormones in our meat and dairy products is to only consume certified organic products or eat less of them.

GMO (Genetically Modified Organisms)

A GMO results from a laboratory process where genes from the DNA of one species are extracted and artificially placed into another species. The result is a new trait in the host species.

There are a variety of different traits that can be passed on to the receiving species. This includes resistance to herbicides and insects, drought tolerance, virus resistance, plus the addition and reduction of various chemicals and nutrients. These traits can be combined so you can have a GMO corn resistant to herbicides, insects, and be drought tolerant.

Did you know?

Most insulin, taken by millions of diabetics to control blood sugar levels, is made with the GMO process.

One of the major concerns raised with GMO foods is around herbicide-resistant GMOs. For instance, a field of herbicide-resistant corn can be sprayed with glyphosate (Roundup©) which will kill the weeds but leaves the corn unaffected. The objection to this is several-fold, but the main points are that the glyphosate gets into the food supply and that excessive use of it could lead to the emergence of superweeds.

There is a strict regulation process in place to control what kinds of GMO foods can be sold. Though there are many GMO foods that are 'approved' in the United States and Canada, the current list of what is produced and sold is short and consists of the following: soybeans, maize (corn), canola, sugar beets, potatoes, apples, papaya, salmon, and squash.

Over the past several decades there has been an aggressive and yes, successful advertising campaign to convince people that ALL GMO products should be avoided. The problem with this is that though there may be some GMOs that are unhealthy with prolonged consumption, there are also some that can be beneficial to you or at least to your pocketbook.

First, the bad news and it revolves around the main concern mentioned earlier of herbicide-resistant crops that have been sprayed with glyphosate. Specifically, for Canada and the United States, this would be corn, soybeans, sugar beets, and canola.

There has been much debate over the last few decades on this and while the FDA has approved the process, saying that the levels of glyphosate in herbicide-resistant crops are too low and if you do ingest any, it will be passed out of your body within a day or two through your urine. However, there have recently been new testing techniques that measure the glyphosate levels in hair and the results are shocking – at least they were to me.

The study that was released in late 2018 was very limited but detected glyphosate at elevated levels in over half of the test samples (130). The reason why this is scary is that glyphosate is believed to be linked to cancer and may also be a driver for antibiotic resistance (131).

The jury on the health impact of glyphosate is still out. Though many studies show that it can be harmful, it has also been used for decades in North America and there have been no dramatic increases in health problems that have been directly attributed to it. The decision is up to you if you want to be concerned about this or not, but my opinion is why take the unnecessary risk?

If you are concerned about your current or past exposure to glyphosate and other chemicals, do an internet search for home test kits for your food or other tests that are available for you.

One key point to note is that it is not the GMO that is dangerous, but the pesticides that can be applied to the GMO. Several GMO products are designed for traits other than herbicide resistance. These GMO products are cheaper than non-GMO alternatives.

GMO Product	Trait
Papayas	Virus resistant
Potato	Virus resistant and modified starch
Squash	Virus resistant
Apple	Extended shelf life
Rice	More nutritional value

Rice and Rice-Based Foods

Rice is one of the common food sources for ½ of the planet and is often a staple dietary item for those living on a tight budget. Due to overconsumption patterns, polluted water, and modern agricultural practices, the thing we need to worry about with rice is **arsenic.**

Arsenic is a nasty chemical that pollutes our food supply, especially in rice and rice-based products. The reason that rice is so susceptible to arsenic contamination is that it needs to be flooded with water for part of the growing cycle. This flooding can expose the plants to arsenic from both the water and from the soil. The arsenic in the water usually comes from industrial pollution and arsenic in the soil could be introduced using pesticides, even if the pesticides were applied decades earlier for other crops.

There is little short-term impact of the consumption of arsenic through rice and foods made from it. However, there are several studies that show real long-term impacts including various types of cancer, high blood pressure, heart disease, and type 2 diabetes. In children, it has been linked to impaired concentration and memory [132].

The concentration of arsenic in rice varies greatly. Testing has found that, in general, white or brown basmati rice grown in California, India, and Pakistan have the lowest levels of arsenic. Limit the amount of food intake that contains rice, such as rice flour, since the level of arsenic is unknown. Also, try to limit the amount of rice-based foods that are fed to babies [133].

Seafood

Whether it be for taste, nutrition variation or availability, many of us have seafood as part of our diet. Having some types of fish in our diet helps to reduce the risk of heart disease, strokes, depression, Alzheimer's disease, and other chronic conditions [134].

The growing issue with some seafood though is the amount of mercury, PCBs, and other toxins that accumulate in it.

The primary sources of mercury pollution are artisanal and small-scale gold mining, coal combustion, production of non-ferrous metals (aluminum, copper, stainless steel), and the production of cement. Though some of this mercury is released into local water systems, most of it is released to the air and is transported hundreds or thousands of miles before coming down to earth in the form of precipitation.

> Mercury is a naturally occurring element, but we produce a lot of it. In fact, we have increased the amount of mercury in the environment by three-fold over naturally occurring mercury sources.

Recent studies have looked at the risk of mercury intake through seafood consumption being reduced if the selenium content of the seafood is high. **Selenium** appears to counteract the negative impacts of mercury and there are even discussions of creating a new mercury-selenium index for seafood.

The accumulation rate of mercury in seafood varies by species, with a general rule that larger and older seafood have higher mercury concentrations.

The health impact of mercury is a bit like the water that many of our fish swim in: murky. It has been established that mercury does negatively impact neurological development fetuses, but in adults, the impact is less certain. To add to the uncertainty, many kinds of seafood also have high concentrations of selenium, which counteracts the impact of mercury.

The table below can be used as a guide if you are concerned about the level of mercury in your diet (135) (136).

Recommended seafood (lower mercury concentrations or high selenium to mercury ratio)	Seafood to be avoided (higher mercury concentrations, lower selenium to mercury ratio)
Rainbow Trout	Shark
Wild Salmon	Swordfish
Oysters	Bluefin Tuna
Pollock	Halibut
Crab	Canned white (Albacore) Tuna
Sardines	
Mussels	
Canned light (Skipjack) Tuna	
Shrimp	

Though the use and manufacturing of PCBs (polychlorinated biphenyl) has been banned in most of the world, there have been many cases where storage sites of old PCB transformers and other contaminated items have leaked into our waterways. Other toxins are typically released into our rivers, lakes, and atmosphere from factories. Though this release is tightly regulated, there are many cases of unintended leakage.

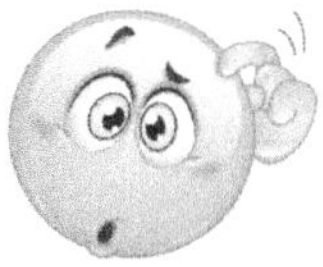

Though PCB levels in India are reaching critical levels, they are still allowed and will not be banned until Dec 31, 2025 (240)!

Since toxin pollution is localized, the best way to find out about the potential hazards for seafood and freshwater fish is through searching for fish consumption advisories in your local region. Using your internet search engine, try the following or similar terms:

"EPA fish consumption advisories"
"Fish consumption advisories California"

Nutrition Content

As discussed in the first section of this book, soil erosion is a serious issue for our ability to grow food and is especially worrisome in Africa and parts of Asia. Our lack of crop rotation also decreases the quality of the soil and the resulting drop in nutritional value of the food grown in it.

Increasing CO_2 values are not helping the nutritional content of some crops either. While plants, to a point, grow better with more CO_2, this increased yield is lower in nutritional content. Take rice as an example, which could see protein content decline by 10% and some vitamin content drop as much as 30% within CO_2 levels that are expected within the next three decades [137] [138] [139].

Keeping up with the proper vitamins and minerals may not be as simple as eating more food. Our food production needs to double by about 2050 and there may not be much to go around. Eating more may also have other health impacts.

One of the better ways to ensure you get the vitamins and minerals that you need is to take supplements. The topic of what the vitamin and mineral content for you is far beyond the scope of this book since it varies greatly between gender, age, and diet. There are several great online resources and books that cover the topic in detail. I have recommended a few at the end of this section.

Food Labeling

There is an exploding demand for food on this planet because of population growth and changing dietary preferences. The industrial food complex is matching this demand through ever-changing techniques in growing and processing food, biotech advances like genetically modified organisms (GMOs), and increasing supply chain efficiency to bring you food from remote regions.

By the time the food reaches your grocery aisle, it will have some mandated labels plus marketing glitz on it, but from this information it is virtually impossible to tell where the product comes from, how it has been processed, what level of contaminants are in it and, in some cases, even what it is.

Again, this is not a book on healthy eating. The labels described below are used more often now as the industrial food complex gets more creative in growing the food we eat and hopefully, this will help you untangle this piece of the food supply maze and narrow down the choices for your specific diet [140] [141].

- **USDA Certified Organic** – Product must have 95% content to meet the organic growing standard. This includes land that has not had pesticides or synthetic fertilizers for three years. In the case of animals, they must be fed organic feed and cannot be given antibiotics or hormones. Farm inspections are done once per year.
- **Certified Naturally Grown** – Same standards as organic but not certified by the national organic program of the USDA. Instead, this is certified by a non-profit organization and farms are inspected by other farmers.
- **Non-GMO Project Verified** – This certification is administered by a non-profit organization and indicates that at least 99% of the product does not contain genetically modified organisms. Validated through inspections.
- **Non-GMO/GE Free** – There are only guidelines by the FDA for this, there is no regulation.
- **Grass Fed** – The animal only ate grass or hay for its entire life. Regulated by producers sending in documentation to USDA's Food Safety Inspection Services (FSIS).
- **Pasture Raised** – The animal must have access to the outdoors for a minimum of 120 days per year. Regulated by producers sending in documentation to USDA's Food Safety Inspection Services (FSIS).
- **Natural/Naturally Raised** – Not regulated.
- **Raised Without Antibiotics** – The animal was not given antibiotics at any time in its life. Regulated by producers

sending in documentation to USDA's Food Safety Inspection Services (FSIS).

- **Raised Without Hormones, No Hormones** – The animal was not given any additional hormones at any time in its life. Regulated by producers sending in documentation to USDA's Food Safety Inspection Services (FSIS).
- **Free Range** – Only regulated for poultry (but does not include eggs) and indicates that the bird spent "some" time outdoors. Regulated by producers sending in documentation to USDA's Food Safety Inspection Services (FSIS).

Diseases and Allergies

Pollen

As we get warmer weather and there is more CO_2 in our atmosphere, there is more plant growth and hence, more pollen lasting over an extended period. If you are allergic to certain types of pollen, this will make your allergic reaction worse. If you currently do not have allergies to pollen, there is a good chance of you developing one as you are exposed to more and more pollen over your lifetime.

Pollen counts could double from today's value in as little as twenty years (223).

There are over-the-counter medications that you can take for this. You can also keep your windows closed, install air filtration that is pollen effective, avoid being outdoors on hot days, wear an air mask when outside or only go out after a rain shower.

Food Allergies

The Center for Disease Control (CDC) has reported that food allergies in children have increased about 50 percent between 1997 and 2011 and are now impacting about 1 in 13 children in the United States (142).

Part of this rise is believed to be with the obsession we have in the Western world with cleanliness such as the use of anti-bacterial soap. Other explanations include the diet of the mother while she is pregnant, a lack of vitamin D, and pollution. The truth is that no definitive answer has been found and even though there are more people with food allergies, this cannot be blamed on the trends discussed in the first section of this book – yet.

Vaccines

Growing entitlement, government distrust, growth of populist political parties, and fact-free news are the trends that are largely behind the "anti-vaccination" movement that is once again gaining strength around the world. The current target for this movement is the measles, mumps, and rubella vaccination which they claim can cause autism in children. Even Donald Trump, prior to becoming president, weighed in on this via Twitter.

"Healthy young child goes to doctor, gets pumped with massive shot of many vaccines, doesn't feel good and changes - AUTISM. Many such cases!"
6:35 a.m. - 28 Mar 2014 - Donald Trump

Of course, when looking at both scientific evidence and long-term statistics, there is no basis for a claim that links autism to the measles vaccination (143).

This anti-vaccination movement is not causing much of a dip yet in the overall immunization rate in the industrialized world, but there are pockets where the MMR (measles, mumps, and rubella) vaccine rate is low, particularly in areas where the anti-vaccination groups are well-organized. In these areas, we are now seeing isolated outbreaks of measles.

What should you do about this? The answer is simple, do not be dumb! Ensure that your children get the recommended vaccinations. Prior to worldwide mass vaccinations for measles beginning in the 1980s, 2.3 million people died each year from measles. This number is now down to less than 100,000 with most of these from developing countries where the vaccination rate is relatively low (144).

Vector- and Rodent-Borne Diseases

> A recent CDC report shows that the number of cases of Lyme Disease has more than doubled in the United States between 2004 and 2016 [224].

Changes in precipitation and temperature along with encroachment on natural habitat are allowing disease-carrying animals and insects to migrate to places that were impossible only a few short decades ago. In addition to the risks imposed by the likes of mosquitos, ticks, and rats, warmer and wetter conditions also allow viruses, bacteria, and waterborne pathogens to thrive.

Combine the accelerant of climate change with increasing world trade and travel, the potential exists for a pandemic like Dengue Fever, West Nile Fever, and even The Plague to crop up in North America.

What can you do to protect yourself?

- If you are out in an area or time when there are mosquitoes, use repellent spray.
- If possible, when out in wilderness areas wear a long sleeve shirt, pants, and a hat to prevent from picking up a tick.
- Tick bites do not itch, at least initially, if you suspect that you have a tick or a bite from one, see a doctor – never try to remove on your own.

If you are interested in the specific risk and potential of risk in your area for diseases like West Nile Virus, Malaria, and Lyme, there are a few online interactive maps that can be found by searching for "online interactive disease map climate change". One example is at this URL:

https://therevelator.org/climate-goes-viral/

Extreme Weather

As this section delves into the various ways that Mother Nature will torture us, suggestions will be given on how to adapt to these conditions if you remain living in the same location. If you are willing to explore more serious measures, such as moving your residence, a chapter will be coming up in a bit on "residence flexibility" and how different areas will be impacted as time progresses.

Extreme Heat

I am guessing that you have experienced at least a few heatwaves over the last few decades, and you are not alone. During this time, we have seen record-shattering heatwaves across the globe causing discomfort, sickness, and even fatalities, especially in areas that are used to cooler weather. Consider the following:

- In 2015, over a **one-week period**, **700 heat-related deaths** were reported in France (145).
- According to the Bureau of Labor and Statistics, between 1992 and 2016, **783 workers were killed on the job** due to excessive heat and 69,374 were seriously injured (146).
- If you think **heatwaves are occurring more often**, you are right. In the 1960s there was an average of two heatwaves per year in major US cities and this **has increased to six per year during the 2010s**. Not only are there more of them, but the season when they occur is also getting longer too, with the season lengthening and is now 45 days longer than in the 1960s (147).

There are several ways that extreme heat can impact you (148).

Impact	Symptoms
Dehydration	Thirst is not always the best indicator of dehydration. For infants, not having a wet diaper in 3 hours, sunken eyes or cheeks, or no tears when crying. For adults, less urine,

	darker urine, fatigue, dizziness or confusion.
Heatstroke	Core body temp of greater than 104 Deg F/40 Deg C. Complete or partial loss of consciousness. Sweating (or lack of) is not a good indicator.
Heat Exhaustion	Heavy sweating, weakness, fatigue, nausea, diarrhea, and muscle cramps.
Heat Fainting	Temporary dizziness and fainting.
Heat Cramps	Sharp muscle pain.
Heat Rash	Small red spots that may itch or have a prickling sensation.
Heat Edema	Heat-induced swelling, usually in the ankles, feet, and hands.

When we are in a hot environment, our bodies first try to cool down by transferring heat to the skin's surface. When that fails to cool us enough, we start to perspire. However, sweating does not work well in conditions of high humidity and is why high heat combined with high humidity is such a deadly combination.

The **heat index** helps us figure out the combination of temperature and humidity and is what the temperature feels like to the human body when considering both relative humidity and temperature.

Most weather services will either report the heat index itself or it can be disguised as the "Feels Like" temperature. If you are having trouble getting the heat index for your area, the site below covers the entire world:

http://www.intellicast.com/

Here is a table of heat index values and the impact that they can have on you (149):

Heat Index	Impact
80-90 Deg F (27-32 Deg C)	Fatigue after prolonged exposure.
90-103 Deg F (32-39 Deg C)	Heatstroke, heat cramps or heat exhaustion possible after prolonged exposure.
103-124 Deg F (39-51 Deg C)	Heat cramps and heat exhaustion likely, heatstroke possible after prolonged exposure.
Over 125 Deg F (Over 51 Deg C)	Heatstroke highly likely.

It is important to note that the heat index reported by your weather service represents the value **if you are in the shade**. If you are in direct sunlight, the value can be as much as 15 Deg F (9 Deg C) more.

Who is most affected? The list is long. Of course, the elderly, infants, and small children are on the list, but we can also add people that have social disadvantages like limited financial resources or lacking cool places or healthcare. Other groups at risk are those that are physically active and those taking medications.

If you take medications regularly, you could be impacted in two ways. The first is a side effect of the medicine that you are taking may be dehydration or make you more sensitive to extreme heat. The other way it may impact you is if the medicine gets too warm and may lose some of its effectiveness [150].

What can you do to reduce the impact of heat if you currently find yourself in a heatwave?

- Keep hydrated, **even if not thirsty**. Wear loose-fitting clothes. When possible, stay out of the direct sunlight. Consider using an umbrella or wide-brimmed hat. If you are sweating a great deal, also consider increasing your salt intake.
- Make sure you alert others if there is a high heat index, especially the elderly.

- Seek a cool environment, like a mall, even if it is just for a few hours.
- Find out if any prescription drugs need to be altered during a heatwave or requires special storage conditions. If you live in an area prone to heatwaves, try to get this information ahead of time by talking to your pharmacist or doctor.
- Draw blinds and curtains to keep direct sunlight out of your home. Eat meals that require little use of the stove and oven.
- Check on loved ones that may have mobility issues.
- If all or part of your work is outdoors, could you avoid working in the afternoon when there is a high heat index?
- If you do not have air conditioning in your home, try sleeping with a damp sheet or drape one over a fan. A bowl of ice in front of the fan can help too.
- Set your ceiling fan to push air "down" into the room, usually counter clockwise.
- Eat spicy foods. The extra sweat will keep you cooler. Eating less salt and protein will also help.
- Make sure you do not leave kids, elderly people or pets in vehicles that do not have the air conditioner running. On a sunny day, if the outdoor temperature is 70 Deg F (21 Deg C), the inside car temp can reach 113 Deg F (45 Deg C) in one hour. If the outdoor temperature is 80 Deg F (27 Deg C), the car can reach 130 Deg F (54 Deg C) in about one hour.
- Take first aid training to be better-prepared for heat-related health incidents.
- If you have a furnace and it is in your basement, turn on just the fan. This will ciruclate cool air from your basement into the rest of your home.

There are also many things you can do around your home to make the temperature more comfortable. These are covered in the topic of "Dealing with Increasing Living Costs" a bit later in this book.

Extreme Cold

From the frying pan into the freezer we go. Though greenhouse gases are warming the planet, they are also causing changes to weather patterns that have resulted in prolonged cold weather events. Analyzing the weather data in the United States has shown that winter storms have increased in both frequency and intensity since the 1950s (151). Of note is the occurrence of polar vortices, where cold arctic air shifts to the south, which seem to be increasing in both length and intensity (152).

Just like the heat index that was discussed earlier, in cold weather there is a wind chill index that can also be disguised as a "Feels Like" temperature during weather broadcasts. Just like the heat index, the wind chill factor can normally be found in your local weather forecast but if not, it can be found at this website:

http://www.intellicast.com/

The following chart outlines how extreme cold can impact you:

Impact	Symptoms
Frost Nip	Red and cold skin. May feel prickling, numbness, tingling or stinging sensations.
Frostbite	Skin appears white or grayish-yellow and may feel unusually firm or waxy. May feel pins and needles, followed by numbness.
Hypothermia	Shivering, slow shallow breathing, drowsiness, confusion or memory loss. Severe cases may result in death.

Windchill Deg F (Deg C)	Impact
32 to 15 (0 to -9)	Low Risk.
16 to -16 (-10 to -27)	Moderate risk.
-17 to -39 (-28 to -39)	High Risk. Exposed skin can freeze in 10 to 30 minutes.
-40 to -52 (-40 to -47)	Very High Risk. Exposed skin can freeze in 5 to 10 minutes.
-53 to -65 (-48 to -54)	Sever Risk. Exposed skin can freeze in 2 to 5 minutes.

Colder than -66 (Colder than -55)	Extreme Risk. Exposed skin can freeze in less than 2 minutes.

What can you do to avoid or reduce the impact of extreme cold weather? Here are some suggestions if you are currently experiencing extreme cold conditions:

- Wear a hat. Forty percent of your body heat can be lost from your head.
- Cover your mouth to protect your lungs from extreme cold.
- Though they may not look fashionable, mittens that are snug at the wrist are better protection than gloves.
- Wear layers of loose-fitting, lightweight, warm clothing. Trapped air between the layers will help insulate you.
- Take first aid training to be better-prepared for cold-related health incidents.

Like extreme heat, there are things that you can do around your home to make the temperature more comfortable. These are covered in the topic of "Dealing with Increasing Living Costs" a bit later in this book.

Hurricanes and Other Flooding events

Unlike extreme heat and cold, flooding, caused by hurricanes, snowmelt or rain, can occur over a very short period of time and may catch you off guard.

The best preventative measure that can be taken for these types of events is to increase your weather awareness by paying attention to weather forecasts at least once per day and to be prepared for emergencies. A whole chapter on emergency readiness is covered a bit later in this book.

Increasing Ultraviolet Radiation (UV)

The amount of UV (ultraviolet) radiation that is making it through the upper ozone layer is increasing and with it, so are our chances of getting skin cancer, cataracts, and weakening immune systems.

Most weather forecasts will give the local "UV" index, which is a measurement of the strength of the UV radiation. This index is linear, meaning that a UV index of 10 is twice as powerful as a value of 5. The index is open-ended. On December 29, 2003, a UV index of 43.3 was detected at Bolivia's Licancabur volcano.

UV Index Value Range	Level	How to protect yourself
0 - 2.99	Low	Wear sunglasses (both in winter and summer!).
3 - 5.99	Moderate	Stay in the shade around midday. Wear sun-protective clothing, UV protective sunglasses, and a hat. Apply 30+ SPF sunscreen every 2 hours.
6 - 7.99	High	Reduce sun exposure. Wear sun-protective clothing, UV protective sunglasses, and a hat. Apply 30+ SPF sunscreen every 2 hours.
8 - 10.99	Very High	Minimize sun exposure. Wear sun-protective clothing, UV protective sunglasses, and a hat. Apply 30+ SPF sunscreen every 2 hours.
11 +	Extreme	Avoid sun exposure. Wear sun-protective clothing, UV protective sunglasses, and a hat. Apply 30+ SPF sunscreen every 2 hours.

The potential impacts from overexposure to UV radiation on your health include the increased likelihood of developing cataracts (hence the importance of wearing sunglasses) and damage to your skin, which is cumulative. Specifically, with your skin, the risks include the following:

- Sun rash on the skin or in worse cases, blistering and swelling
- Wrinkled- or aged-looking skin
- Dark patches
- Precancerous spots
- Skin cancer

Our Pets

Eroding conditions on our planet are not just impacting our health, but are having consequences for Fido and Fluffy, too. Compared to us, they have shorter lifespans and different physical attributes which slightly alter how the trends affect them. The significant trends to be concerned with are pollution, extreme weather, disease, and the evil marketing geniuses for pet food companies.

Extreme Weather

Heat and cold waves pose a serious threat to our furry friends since they cannot tell us if they are getting uncomfortable. They rely on us using common sense and seeing visual indicators of distress. During heatwaves, cats and dogs react differently than us in that they do not perspire and instead rely on panting and sweating through their foot pads to get rid of excess heat. When the cold weather strikes, their different physical nature helps them to *some* degree, but they are still vulnerable.

The following tables give suggestions on how to make your pets more comfortable during extreme weather, signs to look for if they are in distress, and actions you can take if they are impacted.

Tips to reduce the chances of heatstroke
• Common sense! Ensure enough drinking water is available and there is a shady place to stay. • For dogs, a kiddie pool is great for cooling off. • Reduce exercise times and intensity during extreme heat. • Just like human kids, never leave a pet in a parked car on hot days, even warm day can become 'hot' with sun magnification through the glass windows. • Put ice cubes in water bowls to keep it cooler longer. • Trim fur in the springtime. If your dog spends significant time outdoors, **do not shave**, since the fur protects them from the Sun, and it is best to leave at least one inch of fur. • Avoid walking on hot pavement, concrete or sand.
Tips to reduce chances of frostbite and hypothermia
• Common sense! Reduce time outdoors in cold weather.

• Apparel such as boots, scarfs, or sweaters for dogs with thin fur. • Remove ice buildup in paws after walks. • Trim hair around paws to prevent ice buildup. • If they are outdoor pets, provide a warm dry place for them to stay.

Signs of Heatstroke
• Excessive panting • Unwillingness to move or restless behavior • Drooling • Reddening of gums, tongue, and mouth • Vomiting • Diarrhea • Uncoordinated movements • "spoon tongue" in dogs
Signs of Frost Bite and Hypothermia
• Irritated feet • Shivering • Dull mental state • Skin discoloration • Coldness, brittleness or pain-sensitive when touched (especially ears)

Heat Exhaustion/Stroke Treatment
• Spray or apply cool (not cold) water to your pet and then, if possible, apply a fan to maximize the cooling effect. • Frequently offer water to drink, but do not force. • For heatstroke or if your pet does not return to normal after heat exhaustion, make a trip to a veterinary clinic.
Frostbite and Hypothermia Treatment
• Do NOT rub impacted area. • Ensure that your pet is dry and moved to a warm place as soon as possible. • For hypothermia, try warming your pet by wrapping it in a warm, dry towel. • For frostbite, warm the impacted area with WARM water (water that you can comfortably put your hand in).

- Your veterinarian can assess medications, perhaps for pain, are needed for treatment.

Pollution

Many of the activities that you can take to protect your own health from the increasing pollution risk will also help your pets. Regularly changing your air filters, using filtered water, and reducing products in your home with VOCs will benefit Fluffy and Fido too.

There is, however, one additional area you should pay attention to that is related to increased pollution and this is outdoor water sources and letting your pet near them. Blue-green algae, fecal matter, and other water pollutants are on the rise and can cause your pet to become very sick and even result in death. To be as safe as possible, do not let your pets near water that has algae cover or oil/gas slicks, and check for government warnings, either posted at the site or online.

Food

The "Organic" label is one that is highly pushed by pet food marketing companies and is a multi-billion-dollar business with a high-profit margin. As mentioned earlier in this book, **"organic" refers to how food is produced and does not mean that the food has better nutrition**.

There are plenty of good, reasonably priced, non-organic pet foods with good nutritional content. Paying extra for "organic" pet food, in my opinion, is not money well spent. If you are really concerned about your pet and want organic food, consider saving money and making your own pet food.

Diseases

Ticks, fleas, and mosquitoes are thriving in the warmer climate and are increasing both in numbers and in locations where they can be found. This means that we need to be on the lookout for them, not just for ourselves but also for our pets.

For example, cases of heartworm in dogs and cats, which are transmitted by mosquitoes, is on the rise. While this disease used to be seen only in the southern part of the United States, it has now shown up in **all the United States and parts of Canada**. The population of ticks is also expanding along with the reported cases of Lyme disease in dogs.

There are several medications that can be given to your pets that will help to mitigate the risk of your animals getting heartworm. Your veterinarian can give you options that will work best for you and your pet.

Ticks are big enough to detect and usually are found around your pet's head, neck, ears or feet. **After your walks in long grasses or other natural areas, brush your pet** and then run your hands over their skin to feel for any "tick bumps". You can remove these yourself or bring your pet to the veterinary to get the tick removed. Removing them is important since they can carry several diseases, like Lyme Disease.

If you are concerned about potential disease and would like your pet's tick tested, take your animal or the removed tick to your local veterinarian for identification or testing.

Employment and Lifestyle Impact

Redefining Success

Now that we have looked at how we can lessen the mental and physical impact of the globe-altering trends, it is time to start to explore how they will impact the rest of our lives. As you will see, the need to **act and adapt** is just as vital here as in the previous health chapters. That is, if you want to be **successful**. But what does "successful" really mean to you?

For your mental and physical health, the definition of success is easy to describe in that we are successful if we are healthy enough to perform the activities we want or must do. Being successful in our "life", however, is much more subjective. The "life" success goals that we set for ourselves are critical as they are the primary motivators of our lifestyle.

Currently, the measure of success in our society is primarily based on materialism. You are considered successful if you have a nice new car, the latest tech gadgets, the big house, and have lots of money. In today's society, our kids *must* have the latest toys and we must look great through wearing the latest fashions and perhaps even having plastic surgery.

This measurement of success leads people to believe that the more you own, the happier you will be. That things and appearances will make you more popular, fulfilled, powerful, and give you a higher status. However, once our basic needs are met, being "happy" is simply a state of mind; something we choose to be and cannot be achieved. Success is based on goals which are achieved. In other words, happiness cannot be a goal and hence cannot be a measurement of success.

While we still need some materialistic goals for our basic survival, the unhealthy level of "more, more, more" that our society is experiencing today has led us in the exact opposite direction than we desire it to, bringing with it issues such as anxiety, depression, lower self-esteem, and an increase in selfish and narcissistic attitudes.

What does this type of success have to do with the trends? As we will see in upcoming chapters, **the trends, in general, will make life tougher** and, in turn, will make being successful more difficult. That is if we define success **mostly** through the accumulation of wealth, appearance, and material goals or possessions. If, however, we rethink these goals, if we make them more about **what you do** and **who you are** instead of **what you have**, then I believe that even with the impact of the trends, we can be more successful and happier than we are now.

- **Extrinsic Goals** – Goals that are centered around obtaining external signs of self-worth, approval, and validation. Examples are becoming famous, rich, and popular.
- **Intrinsic Goals** – Goals that are centered around your passions, values, and strengths. Examples include strong personal relationships, personal growth, and contribution to others.

Similar to when the need to adapt was brought up in the health sections, I feel it is critical that we also **adapt our perspective of what success means** by shifting many of our aspirations to intrinsic ones. This shift will lead to a sense of wellbeing through having things like good relationships, a meaningful job, and a sense of being part of something bigger. Changing perspective will also drive a shift **from a desire of being material affluent to one wanting to be "time" affluent**. Having more intrinsic goals also means that you can **be successful every day** since they do not have an "arrival" or "endpoint" like most extrinsic goals do.

Before looking at how to adapt, I feel it is important to reflect on how societies today has largely adopted an unhealthy amount of extrinsic goals to identify success. In general, we can break this down into two groups of people.

The first segment is raised this way. Taught from birth that materialism is the end goal, that giving presents show love, and that how you look impacts what people think about you.

The second segment has taken Alice's path from the story of *Alice in Wonderland.*

> Cat: *Where are you going?*
> Alice: *Which way should I go?*
> Cat: *That depends on where you are going.*
> Alice: *I don't know!*
> Cat: *Then it doesn't matter which way you go.*
> - Alice in Wonderland

This group is in a hurry to get moving, to do something, but has no clear purpose or plan in mind. Without a clear purpose or destination, it is easy to get swallowed up with the masses and be influenced by advertising and social influences, thus making them reactive and allowing for the overweighting of extrinsic goals in their lives.

"Efforts and courage are not enough without purpose and direction" – John F. Kennedy

Regardless if the unhealthy amount of extrinsic goals were taught from birth or have come by through external influences like Alice, to redefine how we measure success, we need to reduce the value we place on extrinsic goals such as wealth, social status, power, and possessions.

For our "life" this means having a clear set of goals based on **what we need and who we want to be**. These new goals should drive our behavior in a straightforward direction with a clear purpose. **The challenge is in the execution of defining new values and goals since it involves a major paradigm shift from our current, well-entrenched ways of thinking.**

All of this sounds tough, at least it does to me. It is easy to say, "Okay, I am going to buy less stuff!", but just by doing this are you going to feel more successful? I doubt it. Instead, you will probably feel less successful, since you will still think that "you deserve" these items even though you may not have them. **To start** to make **true change possible**, we need to **reduce external materialistic influences** and **increase our self-esteem and self-compassion**. These two actions will naturally start to shift us towards more intrinsic goals and towards our new definition of success.

Reducing Advertising Exposure

Modern-day advertising is frightening! Having been honed with armies of psychologists, decades of research, and the latest in technology, you really do not stand much of a chance if you are exposed to it. Through social media and the internet, you are presented with advertisements specifically tailored to you, based on your search history, friends, and your demographics. As you watch television, drive down the road or walk through a mall, you are inundated with messages that are designed to convince you to buy things, behave a certain way or alter your opinion.

Based on these factors, if we want to become less materialistic and have fewer extrinsic motivators, we **must avoid** being exposed to advertising and if that fails, we need to develop skills to be less influenced by it.

It would be easy to say to avoid television, radio, the internet, and magazines, but for many of us that would not be practical since we still want entertainment, sports, and news. With this in mind, the suggestions below are ways to still get your "media fix", but with much less advertising exposure to melt your brain. Of course, if you can avoid media altogether, so much the better!

- One quick and partially effective trick to try is simply mute your television when commercials start.
- A highly effective fix for television is to cut the cable and go to streaming services. Most of these are more cost effective

than cable and have no advertising. If you **need** to watch your local 6 o'clock news, purchase an HD antenna to pick up the local stations or pick up the local news from the internet.

- Sports on television is one of the last places where advertising still has a firm grasp. The average NFL game is 3 hours and 11 minutes, of which, on average, 63 minutes are commercials (153). Start recording your favorite game and then take a snooze or have a meal. If you start watching your recording about an hour in and skip the commercials, you will be caught up to real-time by the end of the game.
- Consider getting a device that will allow you to record broadcast television. This again will allow you to skip the commercials.
- Listen to CDs, MP3s or ad-free streaming services instead of live radio or streaming services with advertising. Average radio broadcasts are about 25% commercials (154).
- For the internet browser on your PC, tablet or smartphone, install an ad popup blocker.
- For your internet search engine, check if there is an option to turn off personalized ads. You may still get some ads, but it will be easier to ignore them.
- Avoid newspapers and magazines which are usually 50% advertising – read a book instead.

Reducing Social Media Influences

How often do you do a social media post about doing laundry? Not too often, I bet. Most people post about things that are exciting, a vacation, being with lots of friends or getting a new job. When viewed, it is natural for us to compare our lives to theirs and wonder why we are not doing as well, why we are not keeping up, and why are **they so popular**?

While there are many positive aspects to social media, like most things, too much of a good thing can be harmful and, in this case, may lower your self-esteem. Obsessive competitiveness in trying to get a high number of likes and followers can give you a **false sense of belonging** by being part of groups and having "friends". Another side effect is where your **self-esteem may take a hit** if you are posting tweets that tend to exaggerate how good things are, just to keep up with everyone else.

Below are a few suggestions you may want to consider that will make your social media experience more positive and lower the negative impact it can have on your self-esteem.

- Cut down **your** time on social media. Timebox yourself and then reduce that time each week. There are apps that help you control the amount of time you spend on social media sites (referenced in the chapter on Mental Health).
- With less time on social media, focus on where it is beneficial. Stay in contact with long-distance friends and communicate with like-minded people. For local friends and relatives, visit them in person or talk on the phone instead of keeping up with their social feeds.
- When posting, post things that are interesting to long-distance friends and topics that like-minded friends will find interesting. Do not post, or at least post less often, things that only your local friends would be interested in.
- For a radical move, do not follow your spouse or close family at all on social media. Talk to them instead!
- Do not follow and remove any person on your "friends" list that does not make you feel good about yourself.
- Like your internet browser, review your privacy setting to reduce the amount of information that is shared about you to advertising companies and individuals.
- Make sure that a good portion of your day is **social media-free**.

Increasing Self-Esteem

Several studies have shown that there is an inverse relationship between self-esteem and materialism [155] [156]. To me at least, this makes sense as people with low self-esteem tend to buy more "things" and have more extrinsic goals to validate themselves and feel more secure. The opposite is also true. People with high self-esteem tend to be less materialistic and I believe that we can take advantage of this in our quest to redefine our meaning of success. The following are some ideas that may help in boosting your self-esteem:

- Volunteering.
- Exercise.
- Stop comparing yourself to others (they may appear to be happy but are they?).
- At the beginning of each day, think of one random act of kindness that you can do for someone. Then do it!
- Teach something to someone.
- Just like how you do not want to follow negative people on social media, remove people from your life that make you feel bad about yourself.
- **Be grateful**. It can be life-changing! Take **one minute** out of each day to be thankful for what you have and for who you are.

Increase Self-Compassion

Another way to increase self-esteem is by having self-compassion or in other words, forgiving yourself if you mess up. While it is okay to think about how you could have done something better, dwelling on it does not do any good and this time can be better spent on thinking of how you will do things better now.

Other suggestions

- Declutter (clutter is stressful).
- Budget out of materialism.
- When considering purchasing anything, first look at buying used or repairing. For big purchases, decide what you want

to buy and then wait at least two weeks before making the purchase.

- Do not make shopping a recreational activity.
- Cancel your junk mail.

In Canada

To stop getting unaddressed advertising, put a note on your mailbox stating that you do not wish to receive Canada Post Neighborhood Mail™. Place the note in or on your mailbox or on the inside lip of your community mailbox, group mailbox or postal box [157]. Also, contact the organization listed below to be removed from targeted personalized mail advertising.

Canadian Marketing Association
Do Not Contact Service
1 Concorde Gate Suite 607
Don Mills ON M3C 3N6
CANADA

In the United States

Direct Marketing Association
Mail Preference Service
PO Box 9008
Farmingdale NY 11735-9008
USA

There are also several "opt-out" services you can register with that will send an "opt-out" note to companies that sell mailing lists.

Wrapping up

As you can imagine, the topic of redefining how we measure success is vast and deeply personal and this section has just scratched the surface. Points that are specific to mentoring youth to set good success goals are covered a bit later in this book.

I have listed several books at the end of this section that go into more detail. If you choose to go down this path, then the trends will impact you less since you will be able to adapt easier and still meet **your** goals and, **therefore, be more personally successful**.

Employment and Training

It does not matter if you are currently employed, just entering the job market or are in school training for a career, it is **critical** that you pay attention to where employment trends are heading and not be complacent about your work status. Changes to the employment market over the next few decades due to the **4th Industrial Revolution** will probably have a **bigger impact on the rest of your life than any of the other trends.**

It does not matter if you work in an office, have a trade, or work in the service industry, this revolution will impact employment across all sectors. Do you want to **wait until you are impacted** or would you rather **adapt ahead of time? Remember that somewhere between 1/3 and ½ of American jobs are at risk over the next two decades**[64] [65].

The four major impacts that are the result of the 4th Industrial Revolution include:

- Shifting demand and types of jobs that are available.
- Lack of permanence of work.
- Change in education paradigm to constant learning.
- Shifting locations of work.

Your decisions on how to prepare and adapt to these global employment trends are complicated even further as they must be factored into additional adaptations that you are considering due to the other trends brought up in this book. Another point to consider is that even if you feel your current job is “safe” from this revolution, you may want to consider taking advantage of this upcoming workplace upheaval to get a better job.

To break up this problem a bit, let us start by exploring each of the four impacts.

Shifting Demand and Types of Jobs that are available

There is a plethora of things that will drive the shifting job market over the next couple of decades. Some major drivers include:

- Rapid advancement and integration of technology, like 5G networking, quantum computing, and 3D printing.
- Automation.
- Artificial intelligence.
- Nanotechnology.
- Increasing size of the middle class in Asia.
- Environmental considerations such as changes triggered to combat climate change and pollution.
- Advances in biotechnology.

> The technology is pretty much in place for fully self-driving vehicles, but the laws still must catch up. With 80 billion invested per year [241], this will happen within the next decade or two and there is a good chance that a child born today will never have to learn how to drive.

The 4th Industrial Revolution will also require skills, focusing on integrating technologies and managing the changes that successful companies will go through. Now it is impossible to list all the jobs that will be impacted but the table below lists some characteristics of the workplace that will either trend up or down in employment.

Positive Employment	Negative Employment
Green installation and maintenance. Solar, windmill, charging stations, recycling, etc.	Repetitive manual tasks. Many examples here including more automation in factories, automated cash registers, and harvesting of crops.
Many medical fields, elderly care, pet care.	Fossil fuel related employment. Including exploration, making and maintenance of internal combustion engines, and gas stations.
Tourist-related positions.	Military.
Environmental and sustainability planning and legalities.	Decision-making – stock analysts, loan officers.
Technology-related such as 3D-printing, 5G, and blockchain.	Jobs that the customer can do themselves such as cashiers and banking.
Customer Relation Management, virtual reality, big data, and data analytics.	Para-legal.
Consulting on managing changing technology in a company.	Insurance claim adjusters, autobody mechanics, taxi drivers.
	Data entry, secretaries, telemarketers, fast food cooks.

Lack of Permanence of Work

Freelancers, independent contractors, and flex-workers are all terms used to reference people who are working through non-traditional employment situations and are part of what is commonly called the “Gig Economy”. Currently, 60 million people in the United States work this way and it is forecasted that over 50% of the working population will have this type of arrangement by 2027. Europe is seeing the same kind of trend with roughly 62 million people in this type of arrangement when you combine France, Germany, the United Kingdom, Spain, and Sweden.

What is driving all of this?

- Technology is enabling a better and more robust telecommuting environment.
- The integration and penetration of technology are enabling services ideal for the gig economy such as ride-share, pet care, and yard maintenance services.
- Though still in its infancy, **blockchain technology** will also act as an enabler for the gig economy as it can keep track of exactly who has done a specific piece of work on a project.
- The increase in people willing to do part-time work. This includes those who want to supplement their normal income and those who cannot find their ideal full-time position.

Personally, I have been doing gig work as a contractor for nearly my entire career and it has worked out, but I have seen many cases where it has not. The biggest advantages and disadvantages are:

- Have more control over the type of work and work schedule you undertake.
- Ability to add income by adding in a part-time contract on top of holding a regular job.
- Potential for a better work/life balance.
- No pension, vacation or sick time. For most contract work, the contract can be cancelled at any time.
- Work can be sporadic resulting in unsteady pay schedules.
- Need self-discipline. Produce or perish!
- If you solely do telecommute, there is no commute time and with this the associated saving of time and money. The disadvantage is the difficulty in separating home and work.
- Another advantage if you solely do remote work is that you can work from anywhere.
- You need to invest in your own training and equipment to keep current in your field.
- Chance of social isolation.

- You must always hustle to get your next gig. This usually includes strong networking and may also include advertising.

What types of work are suited to this type of arrangement?

- Pretty much any job that is done 100% of the time in front of a computer and on the phone.
- Ad hoc work such as rideshare, pet care, lawn and snow removal services.
- Project-related work, either done remotely or on-site.
- Temporary workforce placement.
- Several creative positions covering areas like video production, commissioned artwork, and even music collaboration.
- Call centers.
- Consulting if you have specialized knowledge.
- Marketing and sales.

Change in Paradigm to Learning

One of the key points about the 4th Industrial Revolution is to understand that **there will be change**, and lots of it, in many jobs. To be successful, you **must "skill up" frequently**. The days of never going back to school after graduation are over, at least if the lifestyle you want cannot be supported by a minimum- or low-paying work.

Sure, there are exceptions to this rule. The non-impacted jobs listed previously and those that run their own successful company may be able to get away without constantly learning, but most of us must accept the new paradigm that requires us not to be complacent. To be on the lookout for, and take advantage of, key learning opportunities is critical to survival.

This is really where the rubber hits the road for this subject and it is entirely up to you. **No one will be as invested as you are in**

your future and this includes whoever you are working for right now. It is time for most of us to consciously embrace three critical themes:

Evaluation - Education - Adaptation

Evaluation	➢ Find out more about trends and opportunities within your current line of work. Look for websites, organizations, and publications that are related and insightful. From these, see if you can find short- and long-term trends and training opportunities. ➢ Do a regular conscious check of your current skills and training that you have, maybe once every six months or so. Make sure you are on track and correct your course if required. ➢ Examine your current lifestyle and match your work to that. Consider retirement as well, and work that you may want to do then as we are living longer and need to fund our desired lifestyles as we get older.
Education	➢ Learn how to learn. There are several free online classes and books on this subject. Look for courses within your own field. ➢ Look for classes in another field. ➢ Look at general classes. ➢ Log all classes on your resume. ➢ Learn theory. Instead of just learning the mechanics of a position, take the time to get a deep understanding of what is behind it. ➢ Find a good mentor or two.
Adaptation	➢ Accept work out of your comfort zone. ➢ Do not be complacent about your current workplace. Be flexible and accept that your work will probably change.

Shifting Locations of Work

Back in the day, before the rise of technology and globalization, nearly everyone travelled to their place of employment. Work was done locally and things were made locally; much of this has changed and it is up to you to decide if this harms you or if you will adapt and take advantage of it.

The first thing to be explored is remote working or telecommuting. There are many jobs today where people work nearly entirely in front of a computer, on the telephone or take part in online video conferences. These types of positions have the potential to be done remotely, either from people working from home or by having the same job done in a developing country. Call centers, banking, telemarketing, accounting, computer programming, and even many jobs in the medical arena such as x-ray analysis can be done from home or even halfway around the world.

If your firm offers the ability to work from home, take advantage of it if you can. Reduced commuting will save you money and time. If your job could be done remotely but your company does not support this, try to become an advocate for it and get the policy changed. Remote workers save companies money, so it should not be too hard of a sell. Also, working remotely does not have to be all or nothing. Perhaps consider working one or two days a week from home.

If the type of work that you do can be done remotely and your company does not support it and you cannot change their minds, perhaps consider switching to a competitor that does. If you do decide to look for employment, cast a wide net. Many companies will only advertise locally yet are okay with remote work. If your skills lend themselves to it, even look in other countries!

One final thing about remote work. If you do have a job where you work entirely from home and you are living in a high-priced city, consider moving to an area with a lower cost of living.

If you are involved in manufacturing, you need to be open to the idea that your facility may move. With globalization in full force now, it seems that the only way that companies can remain competitive is to outsource some or all of their manufacturing to countries with lower wage costs. For example, the iconic iPhone. The components that go into a single iPhone come from at least 20 different countries.

This does not mean that all manufacturing jobs in industrialized countries will disappear. Through trade arrangements with other countries and the convenience of close-by manufacturing, there are still many sites in industrialized countries. But do not expect there to be a surge in manufacturing types of work and there is most likely little room for wage increases.

The trends will also cause movement of jobs for reasons other than the cost of employment, and this can be used to your advantage if you are open to relocating yourself and your family. For example, if you are a carpenter, electrician or plumber, moving every few years to hurricane-ravaged areas would probably provide steady, good-paying employment. Another example would be to look at small towns that are losing much-needed skills due to urbanization. Though you may not be able to establish a business that you can grow, you may be able to find a niche that provides a steady income with a much lower cost of living.

If you are adventurous, perhaps consider immigration to a different country. Again, this could lead to a lower cost of living, adventure, and different employment and lifestyle opportunities.

Declining Indicators and Stable Jobs

Is your specific job on the path to extinction? If you are complacent and do not pay attention, you can easily find yourself out of work and lacking in marketable skills. What signs should you look for? For many types of work, determining the answers to these questions is complicated since they involve all four factors discussed above.

One way to go about this is to pretend for a moment that you have lost your job and then do a serious look to see what kind of employment opportunities are available to you. If there is a deep market, then you are probably okay for a while, but if there is not a great deal of choice for work that matches your desired lifestyle, then it may be time to "skill up" to be better protected in case of unexpected job loss.

The key for you is to take an honest look at your current or planned career and evaluate the chances of it being impacted, or if there are opportunities that you would like to take advantage of. While every situation is unique, the following are some key warning signs that your current position may be at risk:

- If people leave for any reason, they are not replaced.
- Expenses are more closely scrutinized, less training, and other cost-cutting measures.
- Hiring freezes.
- No cost-of-living wage increases, no or smaller bonuses.
- Machines or software brought in to replace work that is related to your position.
- Opening or expansion of new offices and facilities in other regions while yours is contracting.
- If your company is listed on a stock market and there is a long-term decline in stock price in your industry sector.

While there will be decades of turmoil in the job market, there are many positions that will not be impacted very much. The list below outlines some of these. If you are in one of these positions, even though you may be relatively "safe", you may want to consider taking advantage of this upheaval to get a better job with either more pay or more enjoyment.

- Construction
- Hairdressers, beauty salons, lawn care, pet services
- Healthcare and social workers
- Grade school teachers

A common myth about the 4th Industrial Revolution that is propagated in many sources is that 65% or 85% of the new jobs in 2030 are in "new" professions. After extensive research, I have found no basis for these numbers. Yes, there will be new professions, but probably not anywhere close to 65%.

Summary

Adaptability and anticipation are key. The employment needs of companies over the next few decades will shift quickly and workers that can adapt will have the best chances for success. Unlike the first three revolutions, this time, many good-paying jobs will disappear and if you are not ready for this, you run the chance of only getting lower-paying positions. This means paying attention to trends and constantly training as those that are adaptable, will most likely do better.

If you are considering training, moving or changing jobs, you need to factor in that many of these activities will take time, and this is more reason to start sooner rather than later. For example, it can take months to years to retrain for a different job, especially if you are working at the same time.

When thinking about employment, what you must consider is the type of lifestyle you have and what you want in the future. Will the amount of money you make, the opportunity for raises, and the long-term stability of your employment support what you want or need? Is the type of employment you have something that you enjoy or at least not loath?

There are a few suggestions of good books for further reading on job retraining and the 4th Industrial Revolution at the end of this section.

Dealing with Increasing Living Costs

Over time, all of us expect the prices of most things to rise. The chocolate bar that cost a few pennies a hundred years ago now costs more than a dollar. A cost of a good car in the 1960s was around $3,500, yet today a reasonable one is closer to $25,000. We expect these increases because we accept that we will have inflation, both in the cost of the things we buy and in our wages.

While we expect to have inflation, few of us give much thought as to why we have it. After all, it is a complicated subject with many fluctuating factors such as economic growth, money supply, cost of labor, and the supply and demand of goods. A recent new addition to this complex equation, and one that is often overlooked is the **rising cost of trying to stabilize or deal with sustainability issues**.

A simple example is the worldwide push to stop the usage of plastic straws which pollute our environment. Most of us still like to use straws, so the push is to replace plastic versions with paper ones that cost a couple more cents [(158)]; a cost that is passed from the manufacturer down to you. A more dramatic example is the cost of insurance, which is increasing, in part, because of the increased risk and payouts due to damage caused by more, and extreme, natural events such as frequent flooding, hurricanes, and forest fires.

Though we are starting to see price increases due to sustainability stabilization costs, many of these costs are unwisely being kicked down the road to be paid later, just like many underfunded medical and pension plans are today.

Why does any of this matter? After all, inflation has been around for decades and our lifestyle has remained the same or perhaps has even gotten better. The difference here is that **these cost increases are new and are on top of the existing factors**, they will continue to increase, and are occurring when there is not a great deal of wage increase pressure.

The following sections outline where we are seeing increased costs either due to or for the stabilization of sustainability issues now and where future costs will show up. This information will give you a better understanding of what is to come and, where possible, on how to prepare and minimize the impact on you. **These two things, knowledge and preparation, are key tenants to lowering anxiety**.

The Cost of Healthy Living

In the previous chapter on health impacts, there are many examples where there is a direct cost to you if you want to remain healthy. For example, you either need to buy water or invest in a water filtration system if you want to ensure clean water for personal use. Do you want to avoid particulates in the air from forest fires? You now need to purchase more expensive air filters to do so.

Cost of Carbon

Regardless of your viewpoint on climate change, there is significant momentum for governments around the world to do something to reduce the use of carbon-based fuels like coal, gasoline, diesel fuel, and even natural gas. One way that governments will do this is by making these fuels more expensive and hence less competitive. There are several ways this can be done but the result will be the same: carbon-based fuel prices will increase.

How high will the price of carbon get? This is a bit hard to predict, but an equivalent price of $50/tonne in the next 5 years and perhaps $100/tonne in the next decade are conservative forecasts. Some organizations are suggesting $300/tonne or more by 2030 (159). The table below shows how much these prices would increase the cost of various carbon-based fuels (160).

	CO_2 Price / tonne (US Dollar)		
	$50	$100	$300
Gasoline (US Gallon)	0.44	0.88	2.64
Diesel (US gallon)	0.51	1.02	3.06
Natural Gas (ft3)	2.66	5.32	15.96

Coal (short ton)	105/short ton	210	620

Again, I want to emphasize that **these price increases will occur regardless of the current political climate, where you live or your own personal viewpoint on climate change. The increase may not be a direct carbon tax but may come about through regulations and subsidies as well.**

While you cannot control the price of these fuels, what you can do is ensure that you minimize the usage of them. Listed below are some suggestions that you may want to consider:

- Driving more fuel-efficient vehicles is an obvious choice.
- Live in the smallest living space that you need.
- Taking a fuel efficiency driving course.
- Reduce excess weight in vehicles and regularly check tire pressure.
- Give yourself more time to get to your destination and drive a bit slower.
- Reducing your commute distance (making your commute distance a prominent decision point when deciding where to live).

More suggestions are listed in my books, *Feasible Planet* and *Creation Tithing*. Details at the end of this section.

Increasing Taxes, Decreasing Services

With our increasingly unworkable political systems and ballooning debt, we are pushing burdens that will need to be dealt with in the not-too-distant future. It is simply not politically popular to tell people the true costs of services and it is too easy to kick the can down the road. Combine this with the demographic bubble showing that fewer people will be paying taxes in the next few decades, and it is not rocket science to see the train wreck that will happen.

The impact on you will be less take-home pay and increased expenses if you want to maintain the same level of services that you have today. This can take many forms, examples include:

- Increased taxes at the city, state, and federal level.
- More or higher "user pay" services and fees.
- Less social security income or a higher age to qualify.
- Less maintenance on infrastructure, such as roadways, bridges, and parks.
- Increased cost of goods because businesses will be taxed more, too.

Insurance Costs

The frequency of natural disasters such as flooding, forest fires, and extreme storms has been on the increase for the past few decades and this trend is expected to only get worse as the impacts of the trends increase (161).

Year	Worldwide Major Floods, Storms, Extreme Temperature, Droughts, Forest Fires
1986	260
1996	410
2006	410
2016	700

With increasing populations and people flocking to disaster-prone areas, the total cost of these disasters is staggering. In 2017, the cost of remediating these disasters in the United States alone was $306 billion, and billion-dollar disasters are becoming more common according to NOAA (162).

For most of us, the first line of defense against disasters is property insurance, but insurance companies are in business to make money, so increased payouts and the forecast of the trend only getting worse will result in higher premiums for you. There are several things that you can do to make sure you are appropriately covered and that you get the best deal you can.

- First and foremost is to **understand your policy**. Determine what is covered and what is not. May seem obvious, but get your dang policy out, read it, and make sure you understand each section. If there is something you do not understand, ask your insurance broker or the company that you bought the policy from.
- Consider moving to an area that is less prone to natural disasters. This could simply mean to move to a home that is not in a forested area or a more dramatic move away from a coastline.
- If you live in an area that is prone to disasters, ensure that you have adequate insurance to rebuild as the likelihood of making a claim is increasing.
- If you live in a high-risk area, shop around not only with different private insurance companies but also see what types of government insurance may be available.
- Find out if you can get a discounted rate if you do some disaster-proofing to your property. For example, if you live in an area prone to flooding, you could maybe install a storm surge sewer backup water protector. Another example is installing hail-resistant roofing if you are in an area, like me, that is subject to those nasty ice pellets.
- Keep accurate records of what you own.

Property loss is not the only insurance on the rise because of the unsustainable trends. Health costs are also on the increase due to more pollution and our changing environment. In many countries, healthcare is a social program that is run by the government and hence, you will not have direct control over the fee that you pay, but if you live in an area where you need to buy your own insurance, like the United States, then increases in premiums will directly impact you.

Staying healthy is obviously one of the best ways of reducing medical insurance costs and will also allow you to work longer instead of getting chronic diseases like heart disease, asthma and other lung conditions. Exercise, a big component of healthy living, will help in reducing your anxiety too.

There are many great resources out there on starting to live healthier, a few of which are listed in the Further Study chapter at the end of this section.

Food

Food prices are driven by many factors. Specific to the unsustainable trends, the following will have the biggest impact:

- More droughts, less freshwater.
- Changing climates and biodiversity increases the risk of crop-destroying bugs and disease.
- Growing populations and larger incomes in sizeable countries like India and China are increasing the demand for meat.

Do not go overboard when trying to save money on food by cutting rotten pieces of fruits and vegetables and eating the rest. The "good" portion of the food probably will contain unhealthy bacteria as these goods are heavily fluid and bacteria circulate through the entire piece. Cutting mold off hard cheese is okay though!

One of the best ways to reduce your food costs is to **stop throwing it out!** In North America, roughly 1/3 of the food we buy is thrown out due to it going bad [163]. The best way to reduce food wastage is to do **meal planning**.

Other ways include:

- Avoid at least some ready-to-eat food. A simple example is to buy rolled oats and make it for breakfast instead of eating expensive boxed cereal.
- Buy organics only when it makes sense (as mentioned in the earlier section on our food supply).
- Buying food when it is in season and then storing it also can save substantial amounts of money. Look at freezing, canning, and dehydrating as practical ways to preserve your favorite seasonal fruits and vegetables.
- Simplify the foods you buy. Buy food that is local to the area and less expensive exotic or imported food.
- Learn how to cook. You need not make all your own meals, but making some, like lunch, can save you a great deal of money.
- Eat less meat and more vegetables. Eat an 8-ounce steak instead of a 10-ounce one.

The Cost of Keeping Your Home a Comfy Temperature

As extreme heat and cold weather become more common, the cost of keeping your home a comfortable temperature is sure to rise. The following suggestions may help save you more than a few dollars:

- Obviously, the smaller the area of your home that you are heating or cooling, the better. So close doors and registers to rooms that are not being used.
- Besides the standard lecture of making sure there are no air leaks around doors and windows and having adequate insulation, consider tinting for windows that are exposed to a great deal of sun. With a bit of patience, you can install it yourself, it is not too expensive and can make a real difference.
- Unless you live in a home with zoned air conditioning, you may want to consider getting a small portable room air conditioner unit to keep a few critical rooms, like your bedrooms, cool while the rest of the house is kept at a warmer temperature. The same applies for cold weather and using a small space heater if only a small portion of the home needs to be warmed up.
- Going to be living in the same home for a while? If your lot permits it, consider planting a tree or two that will provide shade for your home.
- If you don't have one already, **get a programmable or smart thermostat and <u>use it</u>.** There is no sense in cooling and heating your home if you are not there. The programmable ones will work well if you have the patience to set them up and keep them updated. Smart thermostats will learn your habits and after just a few days of manually adjusting the temperature, it will take over. The nice part about the smart thermostat is that as schedules change, all you need to do is adjust the temperature manually and the thermostat will adapt to the new schedule.

Fresh Water Cost

In addition to getting scarcer, the price of water will also have upward momentum as governments try to pay for replacing aging infrastructure and keep up with new quality standards. The following are a few suggestions on how to reduce your water consumption:

- In a typical home, the biggest water user is the toilet. Older models use anywhere between 4 and 7 gallons per flush. The newer models use as little as one gallon and still get the job done. Installing a low flush or dual flush model will pay off quickly, usually within a year or two. If you have one of the older models and cannot afford to put in a low flush model right now, consider putting a brick in the tank to reduce the amount of water it can hold.
- Replace lawns with landscaping that does not require watering. You may want to see if your town offers incentives for putting gin drought friendly landscaping.
- A common water waster is leaks. Finding and fixing dripping taps is easy, but **people often overlook the toilet**. To check, put a couple of drops of food coloring in the tank and then check the bowl in about 15 minutes. If the water in the bowl has any of the color that you put in the tank, then you have a leak.
- Another place where you can save a bit of water is by installing low-flow faucet and showerheads. Normal shower heads can use up to 2.5 gallons (9 liters) per minute, whereas low-flow types use 2 gallons (7.5 liters) or fewer per minute. For a ten-minute shower that is a saving of at least 5 gallons (19 liters).

  Replacing the tip of the faucets in the kitchen and bathrooms with efficient aerating ones can also save water. These are easy to put on and come in a variety of flow rates. To determine if it makes sense to change your faucet tip, you need to know the flow rate of your current one and this can be a bit of a challenge. An easy way to do this is by using the chart below after measuring the time it takes to fill a two-cup measuring container.

Time to get 2 cups (0.5 liters) (Seconds)	Flow rate (Gallons / Liters per Minute)
1	7.5 / 28.4
2	3.8 / 14.4
3	2.5 / 9.5

4	1.9 / 7.2
5	1.5 / 5.7
6	1.3 / 4.9
7	1.1 / 4.2
8	0.9 / 3.4
9	0.8 / 3.0

If your current faucet is flowing over 1.5 gallons (5.7 liters) per minute, it probably should be replaced to save you money on "wasted" water use.

Residence Flexibility and Urbanization

The one thing certain about the unsustainable trends is that no one knows exactly how things will play out over the next few decades. **"Not knowing" is one of the major factors of ecological anxiety**.

There are simply too many variables and unknowns, but we do know that there will be significant change that will impact us. Your ability to adapt to change will greatly depend on your choice of housing location and the type of residence where you live.

Life, on its own, will drive different housing needs for yourself and your family. When you are just out of school, pretty much anything will do. When you have kids, more room is needed and when they move out, less room is needed. And as you age, things like stairs are less desirable.

The unsustainable trends add a new set of factors that also need to be considered:

- Increasing population and size of urban centers make it harder to live close to where you work.
- Changing employment opportunities makes for increasing the likelihood of changing jobs, and therefore commute distances.
- Natural disasters that result from the trends, such as flooding, may force you to move.
- Heating and cooling costs will increase because of higher fossil fuel prices and more intense heatwaves.

To remain as flexible as possible with upcoming changes, renting a residence is not a horrible option.

- While some may think renting is just throwing money away, homeownership has interest payments, property taxes, maintenance, and other expenses that are "thrown away" as well.

- Purchasing and selling a home also involves closing costs and realtor fees, meaning that you should stay in your home for a while to spread these expenses out. This may lead you to buying too much or too little house for your current needs. For instance, buying a home with an extra bedroom because you plan on having a child in five years.
- Back to the point of flexibility. Life throws us lots of curve balls and the trends will throw us more. Changing jobs, losing your job, divorce, increased interest rates, and unplanned extra kids are some examples that may force you to sell your home sooner than you would have liked to.

If you are set on purchasing a home, depending on your situation, consider the following:

- If you are planning for children soon and buy a home with an extra bedroom, though privacy is nice, consider getting a border until the children arrive. Do the same when children move out.
- Contrary to modern-day trends, kids do not need their own bedroom. Siblings of the same sex can easily share a bedroom and this will teach them important life skills at the same time.
- Though it depends on your family dynamics, consider the possibility of having retired parents live with you or living with your parents longer than you initially plan to in order to reduce costs all the way around.

If you would like to, or need to move, do not limit yourself to the familiar and perhaps consider moving to a different part of the country or even to a different country. Moving can be treated as an opportunity to not only get a different residence but also to improve your situation with respect to pollution, employment, and expenses.

But where to move to? We have all heard of dire predictions from trends for way out in the future, like in 2100, but what about a more pragmatic outlook, say in the next few decades? What are the best places to move to right now and what factors should influence your decision?

- The one thing to avoid would be extreme weather, something that is getting quickly worse. Specifically, this would include heatwaves, drought, flood-prone areas, and hurricanes.
- Living in or near forested areas should be ruled out with the increased risk of forest fires.
- Living near a great supply of **freshwater** should be a priority.
- The area that you move to should be economically stable. A specific example here would be to avoid states like Illinois, Kentucky, and New Jersey who are way behind in funding pension funds (164).

When considering the above factors and including political stability, wealth, and the availability of affordable social programs such as healthcare, there are few countries that fit the bill with Canada, New Zealand, and several North European countries making the cut. No, the United States as a whole is not on this list because of several areas being impacted by the factors listed above and the wide range of wealth inequality (165).

Within the United States, there are some areas that will fare much better than others including the following list of cities, along with their surrounding areas (166) (167):

- Portland, Oregon
- Seattle, Washington
- Cincinnati, Ohio
- Tulsa, Oklahoma
- Boulder, Colorado
- Minneapolis-St-Paul, Minnesota
- Pittsburgh, Pennsylvania
- Boise, Idaho
- Charlotte, North Carolina

- ➢ Billings, Montana

Urbanization

Expansion of the industrial food complex, outsourcing of jobs, and relocation of factories are some reasons brought up in the first section of this book that will drive to increased urbanization in the next few decades (168) (169) (170). For those of you that live in North America, moving to an urban location will only impact 5-10% of you, but if you are impacted, the lifestyle change will be significant.

Region	% of population projected to be living in urban areas (171)		
	2018	2030	2050
Canada	81	83	87
The United States	83	85	89
Worldwide	55	60	68

While most of us live in an urban setting and do not want to move to a rural or smaller population centers, there is a definite price to pay for this lifestyle.

- ➢ More pollution, specifically air pollution in urban centers.
- ➢ Cities are hotter! Due to more energy consumption and less heat reflected, cities can be 1 to 2 Deg F (0.6 to 1.3 Deg C) warmer than the surrounding area (172).
- ➢ A tendency to consume more ready-to-eat food and manufactured foods with high salt and sugar content. This can lead to increased obesity and other health issues.
- ➢ In general, a more sedentary lifestyle.
- ➢ As urban areas increase in size, more are built in areas prone to natural disasters such as forest fires and flooding.

If you already live in an urban area or are considering moving to one, there are things that you can do to combat these impacts.

- ➢ Get out and exercise more.

- Learn to cook, not just heat your food.
- Follow the recommendations brought up in the "Physical Health Impact" section of this book.

Of course, there is always the option for you of decentralization and moving to a smaller town or rural environment. Not for everyone, but with telecommuting becoming more popular, it could be a great option for you and may be more affordable.

Wealth Management

You may have already guessed that based on the impact of the trends in the first section of the book, things will **start costing you more** while **you may be making less**. This will make an already tense topic, wealth management, even more stressful for you.

Living with Less

In the "Redefining Success" section, the point that was stressed is that most of the true goals in life are probably not materialistic in nature. This is a good thing, too, since having less stuff will save you money!

Even without considering the impacts of the trends, living with less is something that many of us could benefit from. Nearly 80% of U.S. workers report that they live paycheck to paycheck [173], with many not being able to handle a $1,000 emergency without going further into debt. The trends may already be negatively impacting this situation, however, even if in your case it is not, having a concentrated effort on spending less, buying less stuff, and buying smaller used stuff, will improve your financial situation.

I am not saying that each of us must become minimalists, but I think that nearly all of us can agree that we have too much "stuff". "Stuff" that we have paid for and that we need to have a place for. In each purchase that you make, the key question to ask yourself is do you really need it and if so, is there a better, lower priced alternative?

Our daily expenses are another area to look at when living with less. Having a budget that you **stick to** is a start, along with avoiding the use of credit for daily expenses.

Do not stop when you have a balanced budget, continue shaving and cutting costs so you can build up an emergency fund or pay off big-ticket items quicker.

There are many areas that you can consider when building a budget and cutting expenses. The book, *How to Stop Living Paycheck to*

Paycheck, is an excellent starting point (more details on this book are in the "Further Study" chapter at the end of this section). Some outside-of-the-box ideas include:

- Does every driver in your household **really need** their own vehicle?
- Do you **need** to buy a home where each child has their own bedroom?
- If you live in the city with good public transit, do you even **need** a vehicle?
- If the cost of living high where you are at, is moving to a location with a lower cost area an option?
- Do you **need** cable television? Can you get by with a television antenna and a streaming service?
- With free Wi-Fi available in many locations, do you really **need** a data plan on your cell phone?
- To reduce your grocery and restaurant bills, would you consider eating less or no meat on a regular basis?

Building an Emergency Fund

Though an emergency fund is something that all of us should have for reasons outside of the sustainability issues, these trends will increase the likelihood of needing to use it. Think of an unexpected job loss due to the 4th Industrial Revolution or unplanned medical expenses or property damage because of extreme weather.

Building up or already having an emergency fund will reduce your stress and ecological anxiety levels in two ways. First is that right now, before an emergency occurs, you will **know** that you are a bit better prepared for one. Secondly, if one does occur, this type of backstop **will help to ensure that unexpected bills will not add more stress** to what is an already stressful situation.

Detailed financial planning is beyond the scope of this book, but some key points on emergency funds include:

- Funds should cover three to six months of mandatory expenses.
- Should not be your credit card.
- Should only be used for true emergencies and no, a super sale on a Caribbean vacation is not an emergency.
- If you do not have an emergency fund, start one! Set a low target of a few hundred dollars and once you have reached that, up your target.
- If you are building up an emergency fund, put "found" money such as a tax refund or work bonus in it.
- Make sure that you can instantly access the money if needed, without a penalty. This means not investing this in term deposits or the stock market.
- **Do not** use all your money to pay down debt before building up an emergency fund. This may sound counterintuitive but having an emergency fund in place will reduce your need to go more in debt if an issue comes up.

For more information on emergency funds, I would again recommend looking at the book, *How to Stop Living Paycheck to Paycheck*.

Investment Risks and Opportunities

In general, investments can be broken down into three categories: productive assets, interest-bearing, and preservation assets. The trends will have an impact on all of these.

Productive assets have long been advocated by the legendary investor, Warrant Buffet. They are investments that generate profits and cash flow. This could be real estate, stocks in a company or even your own business.

Looking at real estate first, you may have your own residence or additional properties. The risk of extreme weather, the ability of buyers to afford homes, and long-term access to freshwater are some new impacts that the trends will create. Consider the following:

- House prices collapsed by 30% in Flint Michigan since lead was discovered in the water system (174). A recent study shows that in eight coastal states, real estate owners have lost $14.1 billion in home value since 2005 due to flood risk (21).

- Several cities, along with the real estate prices within them, have collapsed as major manufacturing jobs have left due to off-shoring of work or automation (175) (176).

- There are several factors such as income and interest rates that determine if people can afford to buy homes. Looking at both the Canadian and U.S. affordability indices (177) (178) show that it is becoming increasingly difficult to afford a home, with the Royal Bank of Canada calling it a crisis. With the 4th Industrial Revolution impact on the job market, there is a reasonable chance that this downward pressure on real estate will continue for some time. People need to live somewhere however, and with a growing population, investing in rental opportunities looks to have good upside potential for the foreseeable future.

For stocks, there are many ways that the trends will have an impact. Consider the following few recent events:

- Pacific Gas and Electric stock price dropped by 50% after a massive forest fire in California in 2018 that was exasperated by excessive heat and drought.

- Excessive debt brought on the financial crisis in 2008, resulting in millions of foreclosures on personal homes and the collapse of major financial institutions like Lehman Brothers.

The following tables outline some industry sectors that will need to contend with the trends in addition to their normal business cycles.

Sector	Type of Impact	Explanation
Construction	Positive	Replacing aging infrastructure, rebuilding due to extreme weather, the building of protection against weather and rising oceans.
Water conservation, treatment, desalination	Positive	As discussed, freshwater is scarce and needed.
Alternative energy	Positive	Infrastructure, utilities focusing on renewables, etc. Be cautious about companies that rely on government subsidies.
Farm equipment, pesticide manufacturers	Positive	Warmer temperatures will increase forests and open new areas for farmland. More bugs due to higher temperatures.
Brick and mortar stores	Negative	Expansion of internet and mega online shopping sites.
Commercial fisheries	Negative	Decrease in supply of fish.
Vineyards, agriculture	Negative	Changing weather patterns, unexpected droughts.
Skiing and other winter resorts	Negative	Warming trends.
Insurance companies	Negative	Unexpected major losses due to extreme weather.

The other two investment categories, interest-bearing and preservation assets, are in **a precarious state** right now because of the trends. The key for these two asset types are the interest rate, inflation rate, and confidence in the economy; all darn hard to predict at the best of times, and the trends will only make it tougher.

The trends will drive spending and behavior changes by governments, businesses, and individuals as we try to repair, adapt, and stop the impact of the trends. The key is how much will be spent and the level of behavior changes. Here are a few things to consider that may impact all investment categories, but especially interest bearing and preservation assets:

- Due to extreme debt values and a growing trade imbalance, there is a chance that the US government will push for a drastic devaluation of the US dollar over the next few years [179].
- Several reports have come out showing that the impacts of climate change could cause either a slowdown or even a crisis in the financial markets [180]. Large infrastructure damage, crop loss because of drought, and abandoned assets are some areas that could lead to large, unexpected losses in one or more industries which could have a ripple effect through the financial world.
- In the upcoming decades, a great deal of money will be spent by governments and businesses on fixing damage caused by extreme weather, strengthening levees, building sea walls, and replacing existing but crumbling infrastructure like roads and bridges. While all these projects are necessary, they will only maintain existing productivity and will not generate new economic growth. This will likely make the debt to GDP ratio higher, which will lead to higher long-term interest rates.
- We are entering into a very strange market where interest rates on about 33% of worldwide government bonds and 15% of all corporate bonds have a **negative yield**. There are many drivers, and no one knows how this is going to play out, but **this trend is not sustainable**.

Given these possible scenarios, it would seem prudent to take the following actions into consideration, along with the usual market drivers when looking at your wealth management actions.

- Holding a small portion of your portfolio, perhaps one half of your emergency fund, in preservation-type assets like precious metals. This is not an investment but should keep its purchasing power in the event of a major financial crisis or black swan event such as large-scale war or a worldwide pandemic.

- Be extra cautious about long-term interest investments and for that matter, mortgages as well. Interest rates have been historically low for quite a while and could go dramatically higher. If you are purchasing a home, make sure that you can afford the payment if interest rates spike and investments in interest-bearing instruments should not be locked in for more than a few years.

If you are looking for more information on investing while taking the trends into account, there are a few excellent resource books listed in the "Further Study" chapter at the end of this section. If you want to look up additional information, enter "Socially Responsible Investing" in your internet search engine.

Continuous Education

I strongly believe that we should never stop learning. The following gives a brief overview of areas to explore that are directly related to improving your odds against the unsustainable trends brought up in the first section of this book.

First Aid Training

Many of the trends discussed, specifically dealing with pollution and climate change can put you and others in unexpected and immediate danger. Having a basic understanding of first aid can make a big difference in the outcome of these types of events.

- Many organizations provide courses in First Aid, CPR (cardiopulmonary resuscitation), and AED (automatic external defibrillator).
- Cheap and sometimes free.
- Available completely online, in-classroom or a combination of both.
- Boosts your confidence during an emergency.
- Makes you more aware of potential risks.
- Have everyone in your family take at least a basic class.

Learn a Second Language

With increasing immigration and technology that allows people to do many jobs from anywhere in the world, the ability to speak a second or even a third language can be a great advantage to you, especially in the following areas:

- Better job opportunities.
- Wider social circle.
- Travelling.
- Your health, as it works that brain a bit more.
- Builds self-confidence.

Some more popular languages to learn, other than English, include:

- Mandarin (the most widely-spoken language in the world)
- Arabic
- Spanish
- Russian
- French

From this list, Spanish is probably the easiest to learn where Mandarin is one of the more difficult languages to master.

Now you may think learning a second language is hard but with technology, it has become much simpler. Television and online videos can be used to listen to the language and there are many apps that you can use to learn and practice speaking. One example is Rosetta Stone©.

Faith Awareness

In the not too distant past, the community that you lived and worked in was pretty much made up of people with the same background and had similar religious beliefs. The introduction of globalization, immigration, and technology has changed this, resulting in regular exposure to people of different spiritual backgrounds.

The following chart shows a percentage breakdown of the population by religious faith (181) (182) (183) (184).

	The United States	Canada	World
Catholic	22.7	23.3	15.9
Protestant and other Christian groups	48.5	57.1	15.9
Muslim	0.9	1.9	23.0
Hindu	0.7	0.9	15.0
Jewish	1.9	0.5	0.2
Buddhist	0.7	0.6	7.0

No affiliation (Agnostic, Atheist, etc.)	22.8	14.3	16.0
Other	1.8	1.4	7.0

The next chapter will get into the details of the social impact of the trends, with one key point being that **<u>your success</u>** depends on the amount of "social currency" that you have. Regardless of what your faith is, or even if you are agnostic or an atheist, a basic understanding of the other denominations will make getting along with others and developing this currency easier.

A start would be an understanding of the major holidays such as Lent, Ramadan, and Diwali. There are several great books out there to get you started in finding out more on various beliefs and these are referenced in the "Further Study" chapter at the end of this section.

Marketable Skills

My evil day job is doing software and solution development for large companies. I have been doing this for decades and what I have found that helps keep me successful is not just my knowledge in this area but other, more basic skills.

An example would be typing or what they call today, keyboarding. The average person types somewhere around 30 to 40 words per minute, I took typing classes in junior high and still average 70 or so words per minute. Now, my job is not centered on typing, but my speed allows me to bang out e-mails and specifications in half the time, giving me more time to do what I am supposed to be doing, which is problem-solving.

Another good example was a contract that I was working a few years ago. There was a pile of data analysis to be done during this project and though it was not part of my job or expected skillset, I did a bit of Microsoft Access work and a few macros that really sped things up and lengthened my contract.

My point here is that, as part of ongoing learning, you do not have to focus on things specific to a career. The 4th Industrial Revolution, and especially the gig economy, favours people who can be flexible in what they do and have a broad skillset.

Putting in a bit of time to learn a new skill can make you more marketable, improve your performance at work, save you money, and perhaps can even be used to make a bit of money as a sideline job.

Examples of skills that may be of benefit for office type work include public speaking, typing classes, standard office software tools like the Microsoft suite, WordPress for website design, programming, tools to edit content like images and video, speed reading, and accounting.

Some examples of skills that may not be directly beneficial for office work but could save you money or be used as a sideline job, or simply as a skill that you can share with friends to build social currency include pet grooming, cosmetics, gardening, music lessons, bartending classes, local history for background in being a tour guide, and cooking classes.

The Trends

Though the first section of this book briefly discusses twenty of the major trends that are impacting us, is this all that we really need to know about them? For some trends, the answer may be yes, but for others, there are a few reasons that you should learn more about them.

The first reason is that a lack of basic understanding of subjects that impact us and, most importantly, that we can act on, **leaves us gullible and ripe to be taken advantage of**. Consider these examples:

- Politicians on the campaign trail saying jobs in the coal mining or manufacturing industries are coming back. While this may be true in isolated cases, if you understand the driving forces of climate change and the 4th Industrial Revolution, you will realize that these industries are in decline and will continue to do so for several more decades.

- You hear on television or on the news that you should not get your kids the MMR vaccination due to the risks of getting autism. Being wary of fake and misleading news, you research this and find that there is clear evidence that this is false [185].
- From multiple sources you hear that climate change is not real or that it is not an issue. Being informed on this trend, you realize that this is not the case. This book summarizes how it is impacting you **right now** and the trends are pointing to much bigger impacts within the next few decades.

The second reason that you should learn more about the trends is if you are in a position of mentoring youth. Part of this mentoring will be able to explain the trends, their impact or correct misinformation that they may have picked up.

While the "information age" is partially responsible for our decreasing lack of concise knowledge, it can also be used to our advantage to rapidly gain a basic understanding as well. As little as a few decades ago, if you wanted to learn specific details on the 4th Industrial Revolution and how it may impact you specifically, you would need to go to a library and spend hours finding research materials to read. Today, this same search can be done in minutes from your local internet device and you have the option of learning from peer-reviewed material through reading, watching videos, listening to podcasts or audiobooks.

The tricky question to ask yourself is which of the trends do we, in our own unique circumstance, need basic information on and how much detail do you really need to know? **The key point here is to focus on topics that you can act on.** What you decide to research is up to you, but I believe that the following few topics are required:

- Climate change
- 4th Industrial Revolution
- Currency and debt
- Unworkable political systems

Other trends are important, but I feel that a basic understanding of these four is mandatory since they are the easiest to be misled in. To gain basic knowledge in these areas, it is not enough to just start paying attention to them in the news as all news servers contain political or moral bias (or both), which slants their take on any given topic.

I feel your best start is some type of structured documentation in either print, video or audio format as they have a more varied viewpoint and usually have a less heavily-weighted bias in one direction or another. Some suggestions for learning more about these trends are in the "Further Study" chapters of this book.

Travel

Whether it is to experience different cultures, escape from the cold months of winter or to simply enjoy roasting marshmallows while camping, nearly all of us look forward to going away on vacation occasionally. Previously, what dictated where you went and how often was primarily the cost and how much time you could be away from work, but now there are other factors to consider.

Many of the topics in the first section of this book are directly responsible for either the degradation of many popular vacation destinations or limiting how much you can travel. Here are a few examples:

- Changing climate and forest fires, particularly on the west coast of North America, have become more frequent and larger. Camping in this area is impacted through fire bans and closure of camping areas, not to mention that it is not much fun to be in smoke-filled air from fires that may be hundreds of miles away.
- How many of you have a truck to pull a camper or own an RV? It does not matter if you are for it or against it, carbon pricing or regulations will be implemented at some point, thus making fuel for travelling (and day-to-day commuting) more expensive.
- Like the beach? Excessive use of fertilizer on farms and warming ocean temperatures are resulting in toxic and stinky algae blooms around many water areas. Florida has been particularly hard hit with red algae and has seen reduced beach-going tourists.
- Spreading of diseases because of milder winters in many parts of the world. Disease outbreaks in travel destinations such as the Zika virus in Central and South America, and Ebola in Africa.
- Destination degrading. Glaciers disappearing, Venice submerging in water, uncomfortable temperatures.

- Rising ocean levels and changing weather patterns are making the impact of hurricanes more dramatic for winter beach getaway destinations.
- Increased tax burdens coupled with a shrinking middle class mean that you may have less money to spend on a vacation.
- Like to ski? Warming trends are reducing the size of ski resorts and the time that they are open. Like boating? Several popular lakes for boating are seeing their water levels receding.
- Political unrest may make travel to certain locations impossible (China, Russia, and the Middle East for example).
- Unpleasant conditions or the complete shutdown of popular tourist destinations due to excessive use (because of a larger worldwide population) and pollution. An example is Thailand with excessive plastic waste on the beaches and in the water due, in part, to tourist "party boats". Another example is where Boracay Island in the Philippines was shut down for 6 months in 2018 for a massive pollution cleanup effort.

The above points are just a few examples of why vacation travelling is more difficult now than it used to be, and the trends point to this getting worse over time.

These factors are effectively out of your control, so if you have the itch to get out and see the world, you should plan on visiting them sooner rather than later.

Here is a list of some popular vacation destinations around the globe that you may want to consider seeing soon as they will be completely gone or severely deteriorated within the next 2 to 3 decades.

Destination	Impacting Trends
Great Barrier Reef, Belize Barrier Reef	Rising ocean temperature
Venice	Rising ocean levels

Glaciers in general	Rising global temperatures
Dead Sea	Rising global temperatures, urbanization
Amazon Rainforest	Deforestation
Maldives	Rising ocean levels
Galapagos Islands	Rising ocean levels and temperatures
Alaska	Rising global temperatures
Mount Kilimanjaro	Deforestation and rising global temperatures
Taj Mahal, India	Air Pollution, over-tourism, climate change
Machu Picchu, Peru	Deforestation, urbanization, over-tourism
Everglades in the United States	Urbanization and rising ocean levels
The Congo Basin	Deforestation, mining, biodiversity loss
Arctic Polar Bears in Northern Canada	Rising global temperatures
Miami	Rising ocean levels, severe storm damage
Bali	Rising ocean levels, severe flood potential
Mexico City	Significant infrastructure issues due to freshwater shortage. Sublimation causing several parts of the city to sink a foot or more per year.
The Great Wall of China	Over-tourism, erosion, pollution
Patagonia, South America	Climate change, pollution, over-tourism
Olympia, Greece	Threat of forest fires, climate change, pollution.
Komodo National Park, Indonesia	Resource depletion, climate change
Madagascar Rain Forest	Deforestation, climate change
Abu Mena, Egypt	Rising waters, erosion, climate change
Seychelles Islands	Rising water, erosion, climate change
The Alps	Rising temperatures, climate change
Cape Town, South Africa, Mexico City	Lack of freshwater
River Cruises in Europe	Dryer summers, lower river levels

Social Impact

External Influences

We all need to live and work within the social fabric that we are part of. This includes the political systems that govern us, the flow of news to keep us informed, and of course, our coexistence with the corporations that live in the commerce atmosphere. You may recognize these as trends that were brought up in the first section of this book. Trends that influence society and have a direct bearing on the degree that each of us can adapt to other trends.

Fake News—Fact-Free News

I do not know if there is anything that can fuel your ecological anxiety quicker than seeing blatant false news or posts on social media. The topic does not really matter and could be climate change, some political item or a health-related matter. If you are passionate about the subject and are unwittingly exposed to blatantly false information propagated by politicians, news agencies or social media posts, you can have feelings of anger, hopelessness, sadness, and depression. All factors that will feed and heighten your anxiety levels.

This misleading news is hard to avoid as well. You can hear stories from your friends, happen upon them in social media or get exposure through normal information channels such as newspapers, magazines, radio, and television.

Fake News poses a risk to you if you do not recognize it as fake and take it at face value. If you see a Facebook post with 700,000 engagements that reports that a doctor for the Center for Disease Control (CDC) has said that the "disastrous" flu shot is causing a deadly flu outbreak (186), it is natural for you to be alarmed and skip getting the shot. Even though reliable sources, including the Mayo Clinic, show that it is in your best interest to get one (187).

Another way that fact-free news can impact you is that it raises doubt about all news that you encounter. Most of us do not have the time to fact-check everything, this means that nearly everything we read and hear, except for factual items like sports scores and stock market numbers, is questionable. This paranoia and doubt also feeds your level of ecological anxiety.

Your children, especially teenagers who spend a considerable amount of time online, are also impacted by fake news since they may have not yet developed the cognitive skills to question if everything that they hear and read is true or not. False information during these formative years can be damaging for the rest of their lives as it sets up untrue beliefs on a wide range of topics.

Fact-free information is not just limited to news. For example, many online reviews for products are not from real customers but are made up to encourage sales of the products. Another example would be "purchased" followers for online accounts.

With fake news being prominent in our world today and with the outlook that it will only get worse, what can you do to protect yourself from it and handle the ecological anxiety that it feeds?

To avoid fake news, you must first learn to identify it. Some fact-free broadcasts are very well-done and hard to detect, but most can be easily identified if you know what to look for. A few pointers are given below, but also consider taking one of the many free online classes on how to identify fake news such as one put out by the University of Michigan (188) or the University of Texas (189). An internet search for "online class fake news identification" will turn up these and several other options.

- If your news is online, check the domain name to ensure that it is not unusual. For instance, abcnews.com is a legitimate news site, abcnews.com.co is not.
- If there are spelling errors, all caps or dramatic use of punctuation, like multiple exclamation points, it is probably not from a legitimate news site.
- Is the story on other reputable sites?

- ➢ Is there a date on the story? Is it current?

One set of tools at your disposal is fact-checking services and their related apps. These services provide unbiased, verified information on a wide range of topics and news items. A few of the more popular ones are listed below:

- ➢ Media Bias – Fact Check (www.mediabiasfactcheck.com)
- ➢ Snopes (www.snopes.com)
- ➢ Fact Check (www.factcheck.org)

You may see a post on social media or have a conversation with someone that makes statements that you know are wrong. Before even starting to debate this, consider if it is even worth it. Many social media posts are simply trolls and debates can be a waste of energy here. For in-person encounters, a simple "I am pretty sure that is wrong" could suffice. Your time may be better-used on positive activities, such as being with family, friends or in nature than on futile arguments that will only increase your ecological and social anxiety levels.

Social Impact from Corporations

As you may recall from "Corporate Personhood" in the first section of this book, I have little liking for the national and international limited liability private corporations in their present form. In my opinion, they simply have too much power and with the primary motivator of making profits, their actions are major drivers of many of our environmental and social issues today.

The primary influencing tools that they have at their disposal are **advertising**, **political contributions**, and **government lobbying**. They use these to drive sales and to influence laws and regulations in their favour. The question of **how you should adapt** to best **protect yourself** against these money-hungry entities can only be answered by evaluating where **you are being exploited** and determining if you have any **control to stop it**.

The first and most obvious impact that will apply to nearly all of us is how they are convincing us to spend money on products and services. This is done primarily through advertising and they have become masters at it with decades of research and piles of money to determine how to best manipulate us. For us to fight this trend and have it impact us less, **we must adapt**.

For the most part, companies rely on your subconscious to make purchasing decisions. Your subconscious works many times faster than conscious thoughts and comes to a decision, though maybe not the right one, based on multiple inputs and emotions. Advertising will lay the seeds for your subconscious to mull over, such as making you think the product will make you happy, be more popular or successful.

Reducing our exposure to advertising was discussed in "Redefining Success" in the "Employment and Lifestyle Impact" chapter earlier in this book. Though I think it is impossible not to be somewhat influenced by advertising, there are certain actions you can take to lessen the impact on you. There are some excellent reference books on doing exactly this in the "Further Study" chapter at the end of this section. Here are a few of the suggestions from them:

- For major purchases such as a car, television or new outfit, go out and find what you want and then wait 30 days before buying it. This will give your conscious mind some time to determine if the purchase is the correct thing to do.
- When you leave to go to a store or mall, make a list ahead of time of the things you need and **stick** to that list.
- For each item on sale that you are considering purchasing, the question to ask is not if you would buy a toaster at $39 or $29, but if you need the toaster at all. Sale prices should not, within reason, drive whether or not you need something, especially on a non-consumable item.
- Do you need to buy new? Check out garage sales and thrift stores before buying new.
- If you need something you will use infrequently, could you perhaps borrow the item from a neighbor? Along the same

lines, if a few of you are each considering buying a high-cost item, could you buy it jointly and share it?

While nearly all of us are impacted by corporations through advertising, the other way that they ***may impact*** you is by trying to **influence your opinion**. The reason that I am saying "may" is because some people, perhaps including you, do not engage themselves in politics nor get very involved when new laws or regulations are being discussed or voted on. If this is you, then you *will not* have increased anxiety levels from corporate influences since the outcome of the event is not important to you. For the context of this book, this is neither bad nor good, it simply is the way you are, and you can probably skip the next few pages and move on to "Building and Using Social Capital".

If you do get involved in politics, elections, and debates on new laws, then you must be very cognizant of the force that corporations have and how they work to influence your opinion and the impact that this can have on your anxiety levels.

There are two approaches that corporations use to **influence the way we think**. The first one is one that you are already familiar with and this is advertising. A few examples include:

- The insane amount of advertising and robocalls that you are exposed to around election time. The money to pay for these ads usually come directly from corporations, the elite that runs them or through political organizations that are funded by them.
- Promotions done by trade associations such as the National Rifle Association or Beef and Dairy Associations. These promotions work to influence your opinion that their products or services are valuable.

Like advertising that is targeted to get you to buy things, the best way not to be influenced by this is to treat it the same way and avoid it. You will not be saving money, but by being less exposed to this fancy marketing, your stress and anxiety levels will be lowered.

The second way that corporations try to get into our heads is by funding "research" groups, polling organizations, and "think tanks" that release information that is then used by news outlets and political parties. This information comes across as very credible with official-sounding sources like the Heartland or Fraser Institute, but is usually incomplete or misleading and is geared to promote whatever it is that the corporations are trying to get you to believe.

The previous section on "Fact-Free News" has a few suggestions on how to detect and avoid this type of influence. However, while avoiding advertising and ignoring fake news is a positive step in reducing your ecological anxiety, this is only one half of the equation. **Many of us still want to be informed** on the progress of certain topics that we are passionate about. The next section will introduce ways to keep informed that will not rot your brain.

Keeping Informed

While fake news is dangerous, the flood of real news that we are now exposed to is equally formidable. The vast amount of information that is generated and available in seconds is mind-numbing. In this age of information, we have become forced to be very selective about the information that we consume. The situation has gotten so bad that to help us (?), computers are also chipping in with filtering algorithms to steer us toward topics that we show an interest in.

While the expanse of instantly accessible knowledge is great in helping us learn about many things, is it really helping in keeping us informed about things that really matter and affect us? I would suggest that it does not. Many people now feel that they need not keep informed on topics such as the trends discussed in the first section of this book or in other areas such as economics, politics, or global events (190) (191).

There are numerous reasons for this apathy, but regardless, the outcome is a growing population that is content to be less informed. The problem with this is that the phrase "ignorance is bliss" only applies if you are not impacted and have no control. **Ignorance of**

topics that we can act on opens us up to be taken advantage of and exploited. It **makes us gullible** and easily influenced by slick slogans and simplistic, misleading explanations on important topics.

Back in the section on "Continuous Education", there was a discussion and some suggestions on getting a basic understanding of the important trends that are impacting us. This knowledge will help in filtering through the news to find relevant and actionable information.

Thankfully, but also, unfortunately, the impact from the trends on us both now – and into the near future – is pretty much set. We **know** that there will be more plastic in our tap water, that our weather patterns will change, and that for most of us, the job market will be in flux. While there is new technology appearing that will help us adapt to the impacts, it rarely appears overnight and takes time to take hold and become available.

Since the **actionable news** on the trends does not change too quickly, at least not daily, there is no need to scour the headlines every day. Instead, you can consume reliable sources of data at your convenience and there are suggestions for websites in the "Further Study" chapter at the end of this section.

One exception to the above is during political campaigns. Here, the "news" is fast, often inaccurate, and impacts you, yet is actionable by voting. My only advice here is to pretty much avoid all political advertising and instead focus on party platforms, past behaviors, and the moral integrity of candidates. **Fact-check everything**, and most importantly, discuss the platforms with someone who is passionate and knowledgeable about the issues.

As news reporting is done by humans, there will be some bias in everything that you read, hear or watch. It is, therefore, critical to get information on topics you are interested in from **multiple sources**. This will give you various viewpoints, but one thing that this will not help with is **your bias**, and this could be a problem. There is little point in putting in time in catching the latest information if all you are doing is looking for points that align with your opinions. **If you really want to challenge yourself, take a bit of time to understand the basics of critical thinking.**

There are many styles of critical thinking, but the essence of each is to determine the right questions to ask yourself on an issue and then how to apply skepticism, analysis, and rational thinking to come up with a less biased, more informed opinion or judgment on an issue. There are a couple of excellent book recommendations in the "Further Study" chapter at the end of this section and even a simple internet search on "critical thinking" will give many good starting points on this.

One worthwhile area where it is beneficial to keep up with the happenings is in your local neighborhood. This is because local news will likely have an impact on you or at least may be actionable. This is getting tougher as local community newspapers shut down, but there may be some options for you such as local newsletters or perhaps an app like Nextdoor, which is discussed below:

	Nextdoor – An app that is in over 18,000 neighborhoods in the United States alone. Get local news posted by neighbors, references for service providers, buy/sell, and other functions.

Building and Using Social Capital

There are several definitions of "Social Capital". The one I like is that it is the value you and others in society get from interpersonal relationships and other social assets. Examples include the strengths and services offered by your community, the size of your physical and online network, your level of satisfaction at work, and how much trust you have in friends and family.

Unfortunately, our social fabric is continually being torn apart. **Increasing levels of entitlement and distrust between ourselves are eroding our social capital, just at a time when we need it most.**

While there are many benefits to a high level of social capital such as being more successful, happier, and healthier, this section will focus on areas where social capital will help you and your loved ones with the trends in the first section of this book. Specifically:

- Building social capital to be more successful at work.
- Expanding your networks to give and get more support in times of need.
- Supporting and using your community.

Workplace Social Capital

As has already been discussed a few times in this book, our work environment is changing. Job security for many is at risk. There are many things that you can do to be successful at work, **such as working hard, being punctual, and training.** Included in this list is ensuring that you build up workplace social capital. Success in this area will allow you to keep your job longer, increase the probability of promotions, increase the likelihood of pay increases, and at a minimum make you more marketable for other jobs.

Here are some suggestions on building up strong workplace social capital:

- ➢ Avoid gossip. It lowers the perception people have regarding your trustworthiness.
- ➢ Increase the size of your workplace network. Be active with building it up.
- ➢ Reciprocity. Help others without the expectation of something in return.
- ➢ On occasion, be humble and ask opinions, even if you are pretty sure you know the answer.
- ➢ Bridge divides amongst co-workers. Be known as a team player.

Expanding your network

Without thinking about it too much, most of us have developed a network of friends, family, and work colleagues. To build up social capital, we need to work actively to strengthen these existing links while continuing to establish new ones. While some of this may happen naturally, better results can be obtained by consciously working at it.

I think we all enjoy having neighbors like a mechanic who we can ask the occasional question or a musician who can entertain at our parties or get music lessons from. These types of people can easily build up social capital because they have skills that many people can use.

Expanding this further, think of the traits of your good friends and neighbors. They probably are friendly, usually in a good mood, offer to help you with tasks, and will do things for you without the expectation of anything in return. These same traits are the ones you want to develop and expand on to increase the size of your social network.

Some suggestions are given below:

- ➢ Welcome a new neighbor. Offer to help them get settled in your community. Introduce them to other neighbors.

- Plan on doing one act of unexpected kindness per day.
- You keep close tabs on your financial capital, so why not your social capital too? Write down or keep track of your personal and business networks, the people you keep in contact with. Use this list to track who you are staying in touch with to strengthen your existing network.
- Volunteer for local causes. This is good for your community and gives you more opportunities to expand your network.
- Engage with people outside your age bracket such as with the youth and elderly in your neighborhood; they have access to networks and information that you may not have.
- With more education, you can contribute more to your community. Take an evening or online course in something you love and that may be useful to share with your friends.
- Help with or set up a newsletter for your neighborhood.
- Lend tools and uncommon kitchen appliances to your neighbors.
- Ensure elders are engaged. Plant an extra tomato plant for them. Scrape their walks. Rake their leaves. Take them shopping. Help them with technology like setting up television remote controls, social media accounts or e-mail. Volunteer at a local seniors' home.
- Ask for help with things, even if you do not really need it. Reciprocate as well. The social bond in task-sharing is huge.
- Take time to sit on your steps, veranda or common area. Get out to meet people.
- Your social media connections can be just as important as your face-to-face ones. Build an online community with the same principles you use for your physical one. It is easy to spend a great deal of time with this, so make sure that you balance this with the rest of your activities. It is better to be active on just one platform if you are pushed for time rather than try to keep up with several.

- Become an active member of professional contact sites like LinkedIn. Encourage your friends to use them as well.

Supporting and Utilizing Your Community

Imagine that some darn robot has taken your job and you are now looking for work. Do you realize that well over half of employment opportunities out there are never advertised [192]? Most open positions are found through word of mouth and this is one advantage of having a bunch of social currency.

Another way to use your social currency is in saving money. There are endless ways that this can happen. Think of car-pooling, sharing skills, sharing tools, childcare, and pet care, as some examples.

With the trends, there is an increased risk of your area being hit by extreme weather, economic downturns or a pollution event. During such an event, it is human nature to help one another. We see examples of this on the news all the time. However, if you have established a strong social currency prior to the event, you will get more assistance for longer since there is an established bond.

Looking at many of the trends, there are many negative health impacts. Building up social currency and being part of a strong community can help with that! **Strong social interaction will reduce anxiety and improve your self-esteem which, in turn, will improve your physical health, too**.

Supporting Local Businesses

You probably have heard the expression, "Shop Local". Why is this important? If we look at the trends, several point to the rich getting richer, the influence large companies have over our governments, and the importance they place in profits over the environment we live in.

By supporting local companies, you are keeping more of the money in the community instead of having it flow through to Wall Street. The owners of local companies are also members of the community and are more inclined to support local programs such as charities and often pollute less since this is where they live.

Some examples of local companies that are easy to support include:

- Car washes, oil changes, and other mechanical work
- Hair and beauty salons
- Restaurants
- Service companies, such as lawn care and pet services

Embracing Your Faith

This book illustrates the prominent issues that you and I are facing and discusses ways that we can adapt, but I have specifically avoided getting into much discussion as to the core reasons of why we are in the pickle that we are. The simple reason is that we have made mistakes. Through our actions and inactions, we have driven the trends to the point where we must adapt.

Admitting that we are making mistakes is tough since we feel that we are doing the best that we can. That we are living to the "Golden Rule" which is to treat others as we wish to be treated. But are we really doing this? I know that I am pretty upset with my parent's generation for leaving their mess for my generation to clean up and yet our current generation is doing the same thing for our children.

Most of us do not take the time to reflect on why we have allowed greed and selfishness to lead us astray. We may work on cleaning up pollution and reducing our carbon footprint, but we do so because of a sense of duty or obligation without addressing the root causes.

To admit to our mistakes, we can no longer just compare our actions to others, we need to admit that we were wrong, should strive to a higher standard, take ownership of what we have done, and we must have the discipline to do better. None of which is easy on our own.

But we need not do this alone, there is a way to get help. All major faiths have the teachings of the "Golden Rule" at their core. If you have a leaning towards a religious faith but do not get too involved with it nor make it a major part of your life, I would highly recommend that you try to do so.

As a Christian, my guide is the Bible. Through the word of God, we learn the importance of placing the needs of others before our own and showing gratitude for everything we receive. We no longer compare ourselves to others and consider our actions "good enough", our comparison is now to God's standards, which we will never attain, but strive to meet.

Becoming a more active member in your faith community will not only strengthen your faith, but you will also quickly broaden your social network with people with similar beliefs.

Growing spiritually will involve work and should not be left solely to others to do for you. While your pastor and faith community can help you in understanding, they cannot replace learnings and reflection you can get from self-study. By taking an active role in your faith studies you may also find that adapting to our changing world becomes much easier to do since it is now **not an obligation but instead is simply the right thing to do.**

Preparing Our Youth

I think most parents want their children to have at least as good of a life as they have had. Based on where the trends appear to be heading, this looks like it will not happen. This is something that you, as a parent or role model for youth, need to come to grips with and until you do, it will be difficult for you to prepare your kids for what is coming.

The upcoming impact of the trends does not mean that our youth will be necessarily less happy in their lives. If you recall in the "Redefining Success" section, you can be thrilled with your life if you set the right success goals, a skill that our youth will need us to teach and mentor them through. Also, though most of the trends will make life tougher, there are opportunities as well. New types of good-paying jobs, strong communities, and great investment spaces will be available to **those that are prepared.**

How well our youth will do depends a great deal on you. Not so much about your individual actions against the trends such as reducing your carbon footprint, but on how you **teach** and **mentor** them on **reducing their fear** of the future and **how to adapt** to the changes that are coming.

What knowledge and behaviors will our youth need to successfully navigate through these challenging decades, and how can you help with their journey into adulthood? Beyond the life skills that you normally help them with, I believe that these additional behaviors and skills will be necessary:

- A **definition of success** that is based on values that are more focused on who they are instead of what they have and reducing their obsession with materialism.
- **Accurate knowledge** of the trends and where they are heading, the importance of **keeping informed,** and the ability to **filter information** to recognize and ignore misinformation, fake news, and advertising.

- **Essential skills** such as critical thinking, communication, and the ability to learn even after out of school. Most importantly, since there are so many unknowns, the ability and desire to rationalize and adapt by being able to **recognize and avoid behaviors that are counter to the trends** and see **adaptations that work with them.**

Teaching these life skills is made even more challenging in that it must be done while, at the same time, using the techniques to **reduce their ecological anxiety** such as the points discussed in the "Mental Health" section of this book.

Instilling Values for Success

If you recall earlier in this book, our measure of success is defined through goals we reach through having a clear direction and purpose.

Purpose = Values + Strengths + Passions + Action

For our youth, they are in a discovery phase of what their passions and strengths are. Where you can play a role in is instilling values that will prepare them for the trends. These values are similar to what was discussed in the "Redefining Success" section, specifically good self-esteem, humility, and gratefulness.

> "Humility is not thinking less of yourself but thinking of yourself less."
>
> - C.S. Lewis

Humility and gratefulness are values that are best passed down by being a role model. Youth watch and learn through our behaviors. Your involvement in the community, advocating businesses and governments to do more, showing humility, demonstrating gratitude, and your tolerance towards different cultures, races, and religions are all observed by these young minds and there is a good chance that they will pick up at least some of these traits.

Gratefulness can also be taught through increasing awareness in your child if they show ungrateful behavior such as indicating that they expected more presents for their birthday or that they never get to have any fun. When these instances come up, consistently point out to your child that this is ungrateful behavior. Remember — the key is to raise awareness, not to shame them by comparing them to others.

Another way to increase gratefulness is to have them think of one or two things that they are grateful for every day, perhaps at supper time, where it can be openly voiced.

Building up self-esteem in the youth in your family circle is also important. This is a large topic in of itself and there are a few good book recommendations at the end of this section. Some simple points to bolster self-esteem include allowing your children to make decisions, giving them chores, not giving insincere praise, and ensuring to spend one-on-one time with them.

Just as important as building up these values in our youth, we must also be diligent to not erode these important values. Be careful to not lecture them on how spoiled they are or compare what they have with others in the world. This is a form of humiliation and can undermine their self-esteem. Spoiling your child with physical things is something else to be avoided as it can give them a false sense of money and reduce their level of gratitude for what they do have.

The way we reward and punish their actions can also have an impact on the importance our youth place on material items. Paying them or giving them gifts for their accomplishments teaches them that material items are the goal. The same goes for punishing them by taking away possessions. Instead of these actions, rewards can be focused on spending more one-on-one time with them or by having a special outing with them and their friends. Bad behavior can be punished by cancelling special outings or requiring them to do additional chores.

Paying children to do chores is common, but is this really instilling the right values in them? Our children need to learn that doing chores is just part of being in a family and helping others. They still need to learn about financial planning, so they should still get an appropriate allowance and perhaps extra money for doing non-chore-related tasks, but not for regular daily routine tasks.

If you have more than one child of the same gender, a good way to build both self-confidence and self-esteem is to have them share a bedroom. This will build up conflict resolution skills and create a closer bond between siblings.

Teaching our youth how advertising works and how to avoid being influenced by it so deeply is also key in reducing the level of importance that they place on material items.

To close on this piece, we can see that we have a lot of work to do and need to be diligent if we want to instill a set of values in our children that will help them in this changing world. The most important tool you have is facing you in the mirror and is the role model that you provide.

Accurate Knowledge

While we are already seeing the impact of the trends on our lives, it is our youth that will face the true brunt of it and as a result, will have to adapt quicker to our changing world. How well they succeed in adapting will depend in part on how well they understand and believe the trends and their consequences.

Because of limitations in budget, time, politics, and political correctness, we cannot rely completely on our education system to provide this knowledge and must take it upon ourselves to ensure that our youth gain the following:

1. An age-appropriate level of understanding of the trends and most importantly, how the trends will impact them, both now and into the future.

2. An interest in knowing more about the trends that will stay with them as they enter adulthood.

A good place to start is by having ample, age-appropriate, learning material available as a part of this home-schooling effort. This includes books, videos, podcasts, games, apps, and magazines. If you are wondering where to start, the "Further Study" chapter at the end of this section has a list of over 30 suggestions with age ranges.

Having information readily available is a great first step in preparing them for the trends. If we want to maximize the benefit from this and generate interest in our youth, then we must also take the next step of engaging with them and showing our own interest. Remember that you are a role model. If they see that you are interested in these topics and in their opinions, there is a much greater chance that their interest will grow as well.

Asking our youth about the trends and engaging in their learning journey will highlight the following:

- Shows them you are interested in them and their opinions.
- Demystifies the broad, ever-damaging facets of global trends.
- Shows that their input is encouraged in this seemingly "adult" discussion.
- Gives the opportunity to find out what they do not know, which you can then focus on in the future. Do not just focus questions on the "popular" trends, make sure you go through all the topics brought up in the first section of this book.
- Gives the opportunity to point out positive aspects of the trends, such as different kinds of jobs and technologies that will be available.
- Gives the opportunity to find out if they have picked up any incorrect information that you should correct.

- Allows you to find out the areas they are showing the most interest in. From here, you may encourage a more active role for them in battling these trends.

Your own experience can also be a valuable teaching aid. Get them involved in the things you already do or are planning on doing to adapt to the trends. Simple examples include getting them involved in budgeting, food choices, explain your ongoing education plan, and community involvement. Bringing up current events that relate to the trends and explaining the relationship can also be a powerful engagement tool.

Essential Skills

Our current education system is geared to the delivery of information and then testing how well it is understood. This can work for topics like math, reading, history, and geography but is not well-geared to developing character and soft skills. Unfortunately, these are exactly the skills that many will need for understanding the world around them, the trends, and for employment due to the 4th Industrial Revolution.

Skills like critical thinking, collaboration, creativity, and design thinking cannot be completely taught out of a textbook but instead are cultivated through guided hands-on experience. A list of key soft skills and a brief definition are in the table below:

Soft Skill	Description
Communication	The ability to communicate a complex topic in a simple way.
Computational Thinking	The ability to break down problems into small pieces, to do analytic thinking on ways to deal with and delegate problems.
Design Thinking	The ability to design something, test it, and refine it.
Collaboration	The ability to work well with others, organize work, support others, take and give instructions.
Creativity	The desire to try new ways of doing things and to not take existing ways as the only or best way of doing things.

Curiosity	The desire to learn more.
Learning how to Learn	The ability and desire to take an unknown subject and learn about it outside of the structured school environment.

Another disturbing trend right now is that while our youth are readily adopting technology, there is a lack of interest in pursuing the underlying topics behind it. These are referred to as STEM topics (Science, Technology, Engineering, and Math) or STEAM topics (a variant of STEM with the addition of Arts).

The mentoring of our youth in the softer skills and in developing an interest in the STEAM topics can be started before they enter school. A starter list of books, games, toys, and websites are provided in the "Further Study" section a bit later in this book. A few starting suggestions include:

- If possible, guide your children into computer science courses and programs. Computer science is not just about programming, but also include topics like computational thinking, artificial intelligence, data analytics, networking, and robotics.
- Have youngsters work together to make up a new and nutritional dessert, an obstacle course for a pet, a play or a song.
- Does your child have trouble getting up in the morning or finishing their homework? Instead of making a new set of rules for them to follow, have **them** come up with solutions to these problems.
- Encourage participation in team activities.
- Ask them open-ended questions.

Wrapping Up

The coming decades will be challenging for our youth and how well they navigate through them will depend a great deal on you. Throughout this book, there has been an introduction to topics where we can mentor them.

- Helping them cope with ecological anxiety.
- Defining success that is less based on material needs.
- Learning about the trends and their impacts.
- Identifying and developing the skills needed to be successful in the 4th Industrial Revolution.

The more you become involved in any or all these areas, the better the chance of success your children will have. Without this type of mentoring, they will be at a disadvantage in coping with, and adapting to, the changes the trends will bring.

Dealing with Differences

If you close your eyes for a moment, can you imagine yourself as a dictator over the entire world? Where everyone on the planet agreed with your opinions and acted accordingly? If anyone was foolish enough to disagree with you, the consequences would be... well, I will let you determine that.

Now, the sad part here is that you must, at some point, open your eyes and realize that this is not the case. The real world has a full spectrum of opinions and you will constantly run into those with beliefs that are different from yours. With each encounter, you run both the risk of heightening your anxiety levels and, since we are reliant on others for many things, hurting your own ability to succeed if you choose to isolate your interactions to only those with similar beliefs to your own.

Tolerance

We are part of the largest and fastest melding of people ever seen by mankind. News from all over is instantly available via the internet. We now work and network with people from all over the globe. Immigration is on the uptick. These trends all result in more exposure to different values, beliefs, and opinions, forcing us to a whole new level of social integration.

That is right, **forcing** us, that is, if you want to be successful. Not too long ago, communities, cities, and even some countries were, by and large, comprised of a single race of people who had a common set of beliefs and views. That time of commonality is pretty much gone and **highly unlikely to come back**. In today's world, your doctor may be an atheist, your boss an African American, your children's teacher may be part of the LGBTQIA community, and your local politician may be Asian.

Though there may be a few exceptions, most of us now rely on a diverse group of people to survive and for our success. The diversity that is brought about by the wide variety of perspectives that this mix of people bring **can** make a community stronger. However, a community can quickly become fractured and weakened if there is discrimination, prejudice or intolerance to this diversity.

> ***Tolerance*** - *The ability or willingness to allow,* ***within reason****,* the *existence, occurrence or practice of something that one dislikes or disagrees with,* ***without interference*** (193).

Being more tolerant does not mean that you must accept the viewpoint, origin or opinions of other people. But being more tolerant allows you to not get as excited and lathered up about someone being different. This should reduce your anxiety levels and has the added benefit of having access to a broader community to help you be more successful.

The next few pages will discuss suggestions on how to increase your tolerance and get a better understanding of the differences we are seeing in the growing mix of people we interact with.

Learning about beliefs, values, and customs of others

An obvious place to learn about people with different backgrounds is through reading. Therefore, it was brought up in a topic earlier in this book labelled, "Ongoing Education". Some time spent on studying the basics of major religions would be an excellent place to start and a few good reading suggestions are in the "Further Study" chapter at the end of this section.

Reading can only take you so far though in understanding different cultures. Getting out to multicultural events, inviting people of different backgrounds over for a meal or attending different religious services are excellent ways to get firsthand experience.

These learning opportunities will not happen "naturally" through your normal day-to-day activities. Make yourself a to-do list and work through it and do not forget to share your new-found knowledge with others!

Holidays are great opportunities to learn about religions and cultures other than your own. With the diversity of our population increasing, understanding the important holidays of others helps build tolerance and community. Here are some holidays that you may want to learn about (do not forget to teach your children about them, too):

Holiday	Culture/Religion
Christmas	Christian
Easter	Christian
Hanukkah	Jewish
Kwanzaa	African American
Advent	Christian
Passover	Jewish
Yom Kippur	Jewish
Ramadan	Islam
Diwali	Hindu
Bandi Chhor Divas	Sikh
Muharram	Islam
Vaisakhi	Sikh
Uposatha	Buddhist
Vesakha	Buddhist

Educate others on your beliefs, values, and customs

Take every opportunity to teach others about your beliefs. This can be done in a variety of ways. Some examples are:

- Teach a skill from your culture such as cooking or games.
- Explaining important cultural or religious holidays.
- Share your music and art with others.
- Share food from your region.
- Share your sports. Invite others to a local game or to your home to watch one.
- Inviting others to your cultural events.
- Be open about your beliefs. Consider the following example:

> Your religion, such as Islam or Mormon, forbids you from gambling. Your office co-workers start an office lottery pool where each person contributes $10 which is then pooled together to buy a bunch of lottery tickets. When coming around to you, instead of saying "no thank you", try saying "No thanks. My religion prevents me from any type of gambling."

Understanding our own biases

Each of us knows if we are biased against a certain sex, race or religion. We may hide this from others due to fear of social backlash, but this does not change how we feel inside. If you wish, you can try to undo this bias, but at least it is a known trait.

But what about **implicit bias**? Where you **think** you are fair to a certain group, but your actions, driven by your subconscious, demonstrate something else.

There is an entire industry around testing and teaching workers, students, and individuals about implicit bias. Many major corporations and learning institutions have regular programs regarding this and millions of people have been tested or attended workshops. But do these tests work? If they do work and identify your subconscious hate for anyone that wears mittens, can you then really do anything about it?

To say that the tests work well, they should yield consistent results each time they are taken. This appears to be a problem though with several studies showing that the consistency of results from repeated tests is low.

If it is determined that you have a hatred toward people who wear mittens, awareness alone may not be enough for you to control it. Decisions and actions are handled through either conscious or subconscious thoughts, and your subconscious mind is much faster. Unlike conscious decision-making, your subconscious does not need all of the facts to come up with a decision. Adapting your subconscious thought patterns takes a great deal of effort, and awareness alone may not be enough to change it.

Proponents of these tests argue that even a slight correlation of test results and awareness of any potential bias is better than nothing at all. I tend to agree with them. If you know of the limitations in advance and treat the results accordingly.

One test to measure your implicit bias is called the "Implicit Association Test" or IAT. This test has been around since 1998 and involves measuring your associations between concepts and attributes. For instance, good or bad, male or female, black or white.

A few other tests include Extrinsic Affective Simon Task (EAST) and Go/No-go Association Task (GNAT).

If you are interested, there are several places online to take the IAT. One of which is at the following internet address:

https://implicit.harvard.edu/implicit/takeatest.html

An alternative to the tests is to discuss implicit bias with a good friend. Someone who knows you well. They may be able to tell you of biases they have observed in you, which may be a more accurate indicator than test results.

Awareness of how we treat others

Our interactions with family members, friends, co-workers, and strangers give us a multitude of opportunities to either *promote* or *handicap* diversity. Being aware and respectful of the opinions and beliefs of people you interact with builds constructive relationships, which is a direct benefit to you. Consider the following suggestions:

- Apply what you have learned about different cultures and religions to avoid insensitivity. For example, if you know your friend is a Buddhist, then encouraging them to gossip shows you do not either understand or respect their religion (lying or gossiping is one of the five moral precepts that are to be refrained from in Buddhism).

- Acknowledge special occasions and events. For example, if you have a gay son, ask them how they will celebrate an upcoming Pride event or if you have a Muslim co-worker, wish them a happy Eid Al-Fitr and Eid Al-Adha at the beginning and end of Ramadan.
- Your respect for others is not bound to the physical world but extends to internet social communities as well. In fact, these are probably even more important than your interactions in the physical world. In person, a biased term may be excused as a simple misspoken word. However, when you type a message, you have the opportunity to review it before posting, so mistakes are less tolerated. Also, your message can be read, re-read, and forwarded for others to analyze.
- Do not treat everyone the same! Treating everyone the same is forcing assimilation; exactly the opposite of what we are trying to achieve with promoting diversity. Embrace the differences while at the same time, treating everyone with an equal level of respect.
- I am sure that most of you have heard the expression "treat others like you would like them to treat you" or some variation of that. If you do not think about it too hard, it seems like good advice. However, if you *really think* about it, this means you would treat everyone the same, regardless of race, religion or culture. Probably not the best advice for showing tolerance.
- Many topics related to beliefs are controversial. When discussing these, do not personalize it and stick to the topic.

Handling Intolerance

So far, this section has focused on actions **you initiate** with others, but what actions should you and I take, if any, when we are exposed to beliefs, opinions, and cultures that do not align with our own? I am not referring to biases that are against the law in the region you live, but beliefs and opinions that are legal, yet still conflict with your own beliefs.

Since we are talking about **your** beliefs and opinions, anything that conflicts with them will probably annoy, irritate or anger you. In other words, it will most likely **offend you**, at least to some degree.

As discussed earlier, with instant worldwide communication and vast migration of populations, the community you interact with is becoming more diverse. Unless you plan on living in isolation for the rest of your life, it will be nearly impossible to avoid being exposed to an increasing number of different beliefs, cultures, and opinions. This will lead to you experiencing more "offensive acts", and how we individually react will decide how strong or fractured our community will be.

Because we are talking about **your beliefs**, it is easy to get emotional when hearing or seeing something that goes against them. I believe the first and most important thing to do is to ensure that you remain in control of your emotions, regardless if the offending act is intentional or not. Why should you let someone, or something, control the way you feel? **Why give them that power?**

If you find you are offended by something that is lawful, you may want to check the level of your Wii-FM (entitlement). Enriched entitlement can lead to the inability to tolerate unpleasant feelings or stressful situations; a condition called Low Frustration Tolerance (LFT). Take a moment to reflect if the action is truly offensive, is it targeted to you, and what the appropriate level of response may be. This internal reflection will be difficult if you are irritated or angry and is one reason why it is important to keep your emotions under control.

Let us review our definition of tolerance, paying specific attention to the last few words:

> ***Tolerance*** - *The ability or willingness to allow,* within reason, the *existence, occurrence or practice of something that one dislikes or disagrees with,* ***without interference*** (193).

"Without interference" does not mean we cannot debate or protest conflicting beliefs, opinions or actions, but it does mean that we must allow others to openly practice or discuss them. Failure to do so will either suppress the beliefs in question or lead to unproductive and perhaps dangerous confrontations. Either of these outcomes will weaken our community, exactly the opposite of what we need for a sustainable society.

Summing it up

You do not have to like people that wear mittens, believe that mittens are cool or feel like you are forced to wear mittens. The main point here is that the more intolerant you are to those darn mitten wearers, the more you are hurting yourself.

To survive and succeed in the upcoming decades, we need the help of our community. We cannot succeed on our own. Our increasing population, and the migration of it due to the trends discussed in the first section of this book, nearly guarantees that our community will become more diverse and it is up to you if you want to take advantage of this or not.

Coping with Deniers

I occasionally volunteer to do roadside cleanup and each time I do, I am amazed at the amount of litter that is found. Thrown out fast food containers and other garbage that some people, who just don't give a damn, have tossed out of the window. I also often meet people, either in person or online, that do not believe that some of the problems outlined in the first section of this book are real and serious or think that they are blown out of proportion. There are even entire political parties who base their platforms on this ideology.

Encountering a person or a group like this can quickly raise your ecological anxiety levels with your brain screaming, "Don't they understand how serious this is?" or "Don't they care that they are impacting everyone, including me?". Another risk is that the denier will cause you to doubt the trends and hence, weaken your resolve to act.

The instinctive response when coming across deniers of climate change, pollution or any of the other major trends is to try to correct them. To wave this book in their face and expect them to understand. To flood them with facts and figures. This approach, however, rarely ends well, with a debate ensuing and no resulting change of opinion of the denier. In fact, it may even be damaging, since you have given an opportunity for the denier to have a voice, which may influence others, cause you to have doubts, and fan the flames of your own eco-anxiety.

One dangerous group of deniers are the bots and trolls that you can encounter while you are in your online neighborhood. These "individuals" are programs or people who are paid to support a certain viewpoint like climate change denial. Their usual purpose is to sow doubt behind the facts and to give "internet" oxygen to whatever it is they are promoting.

For you personally, there is little to be gained by engaging with the bots and trolls, as their opinions cannot be swayed. There is a danger though if their posts start to influence you and hence, it is important to be able to identify if a post is written by one. But identification can be tough. Trolls and bots adapt quickly, so the best way to identify them is with an app such as botsentinel.com. You may be able to find new identifier apps by typing "troll bot detector" in your internet search engine. Social media sites also try to help. For example, Twitter has botometer.org. Here are a few red flags that can alert you if you may be engaged with a bot or troll:

- No profile pictures.
- Lots of numbers in their account name.
- Rapid posts (a second or two apart).
- High frequency of posts (more than 100 per day is suspicious).
- Few followers and piles of "likes" or "retweets".

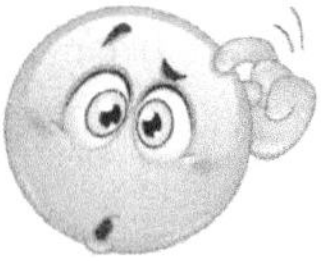

Handling Bots and Trolls

If you do identify a bot or troll, here are a few handy tips to make you less appealing to them, along with making life a bit tougher for them:

- Do not engage. Engaging with them only tells them that they are getting to you. It can also enhance their popularity on the social media platform.
- Do not post that you are being targeted, it will only make you a bigger target for other bots and trolls.
- Always block and if threating or extremely vulgar, report to the social media site.

Now how about the "real" deniers? The ones that you encounter in person or are "real and genuine" social media users. For this group, I believe there are two alternative courses of action that you may want to consider. The first is to not engage at all. With most skeptics, there is little chance of you convincing them to change their minds and a debate will do little to lower your anxiety levels, in fact, it will probably increase them. This does not mean for you to be disrespectful to them, but just to not challenge their point of view.

A good example is with some politicians that are now in power and the climate change denial stance that they have taken. You could let this get to you, let it fester up and create anxiety or you can consciously decide to not engage, at least until election time.

The other approach is to skip the facts and start telling stories related to the impacts. People relate better to storytelling and it brings up points that cannot be refuted. The table below gives a few samples, and I am guessing that you will have a few local examples of your own as well like flooding, drought, job losses or pollution events.

Instead of this fact	Try telling this story
The average temperature in Europe has increased by 1.5 Deg C (2.7 Deg F) over the last century and the frequency of hot days has tripled (194).	Wow, did you hear that last summer, for the first time ever, they had to shut down nuclear reactors in Finland, Sweden, and France because the intake cooling water was too warm?
18 billion pounds of plastic flow into the oceans each year (195).	Hey, did you hear that they did tests on tap water around North America and nearly all of them contained microplastics? This can't be good for us if we drink it.
If we can hold the increase in global temperature to 2 Deg C, then there is still an expected 30% loss of areas with permafrost (196).	Just read an article that said there is a chance that old diseases like the bubonic plague, anthrax, smallpox, and the Spanish flu may come back after being released from the melting permafrost up north. There has been one outbreak in Russia already.

In the U.S., wildfires are burning twice the area when compared to the 1980s and 1990s [197].	I hope this summer will be better than last summer when we had a few months of smoke in the air that was so thick, it reduced visibility to a mile at best.
25% of jobs are at high risk of being automated.	Give a local example. Mine is that there are 5,000 fewer people employed in Alberta's Oil, Gas, and Mining industry now when compared to 2008, despite much higher output of fossil fuels.

Emergency Readiness

Why Now?

All of the issues discussed in the first section of this book are impacting us right now and it seems that things will only get worse in the upcoming decades. Some of these issues, mostly the ones related to weather, are increasing the risk of us encountering real emergencies. Situations that happen with little or no notice, and that may cause hardship to us or our family in one form or another.

On some level, nearly all of us prepare for certain types of emergencies. This can be as simple as carrying a spare tire in your vehicle, signing up for health insurance or just having a hidden stash of coffee beans (yes, running out of coffee is an emergency). There are other types of emergencies that some of us do not plan for, including earthquakes, flooding or even death without having life insurance, even though you have small children in your care.

Why do we plan for some types of emergencies and not others? Cost of protection, potential impact, and the likelihood of occurrence are the major factors that you and I consider when determining what to plan for and where to take our chances. Now, some of the trends will increase the frequency and magnitude of events that we may encounter. A few examples would be:

- More flooding due to changing weather patterns, producing prolonged rainfalls, rapid melt of mountain snowpack or more frequent and stronger hurricanes.
- Prolonged power outages because of overloaded generation systems in hot weather or downed power lines due to flooding, or large snowstorms.
- Disruptions in water supply because of the discovery of unsafe toxin levels due to expanded population, increased manufacturing or aging infrastructure.
- Terrorist acts due to actual or perceived intolerance of a specific group of people.

- Increased number of forest fires due to changing weather patterns.

This is what is different <u>now</u>. What may have seemed okay to take a chance on a few decades ago may now be an area where it is prudent to do at least some preparation for as the likelihood and level of impact are increasing.

How well you fair in any emergency will depend on how much advanced preparation is done and on the quick decisions that you make during the crisis. Lack of planning can result in distractions during the critical situation that will divert some of your focus and hamper your decision-making ability. Let us look at a few examples:

- Not a major emergency, but what if you lose your cell phone while away from home and then need to contact your spouse or friend on an urgent matter? Do you know, from memory, their phone number? Personally, I no longer memorize phone numbers since they are all stored on my phone. A simple credit card-sized piece of paper in your purse or wallet with critical contact numbers could resolve this dilemma.
- A more serious emergency would be if you are at the scene of an accident and there are injured people. You have access to a first aid kit, but do you know how to use any of the contents besides bandages? Introductory first aid and CPR courses can be taken in just a few hours for low cost or free!

Advanced planning will not only reduce the impact of a crisis but can also reduce your stress and fear levels during the emergency. Stress and fear are significantly driven by the unknown and worries about what may occur. Consider the following examples:

- A simple example but if your car breaks down on the way to pick up your child from school, if you and your child have a plan on what they should do under this circumstance, then this is one less thing to worry about beyond getting your car fixed.
- If a significant storm runs through your area and results in you not having water or electricity. You will be scrambling

to get drinking water right away, however, if you have a 72-hour emergency kit in your home, you will have at least 3 days of water.

- You unexpectantly lose your job to a silly robot and will not get severance. If you already have a financial plan in place to keep your household running for a few months, such as a 1-month and 6-month emergency fund, then you will be able to focus more on finding a good job instead of scrambling to get any job just to pay your bills.

Preparation

Having things in order ***before*** shit hits the fan is essential. The following is a list of common items that most of us should consider preparing now while things are calm and you have the time to do them.

Passports

Keeping a valid passport with an expiration at least one year out has always been a good idea and now with sustainability factoring in, there are even more reasons to have one. Consider the following:

- You may not have a crisis, but relatives and others in foreign countries may. Perhaps you need to visit a family member who has suddenly become ill or travel to a country to support family or others who have been impacted by a weather disaster.
- A sudden opportunity comes up to travel for work or vacation. Having a valid passport will allow you to travel on a moment's notice without the added expense of trying to get a new one through an expedited process.
- Lost your driver's license or state identification card? It can be replaced but may take a few weeks to do. A valid passport serves as an excellent backup for photo identification.

Your Smartphone

For most of us, our cell phones are never very far away and during an emergency they can serve a dual-purpose. The first is to call for help. While in the United States or Canada, dialing 911 will get you in touch with emergency services. If you are in another country, dial 112 instead. **Nearly all phones will allow you to dial 911 (or 112) even if it does not have a SIM card.**

The other use for your cell phone during an emergency is to speak for you if you are incapacitated by telling others your "In Case of Emergency" (ICE) contact list and perhaps even critical medical information. This ICE information can and should be available to others, even if your phone is "locked" and nearly all newer smartphones have this capability built-in. If they do not, there are apps that you can download for this function. If you are like me and have issues setting up your phone, reach out to a tech-savvy friend to help you.

Contact Lists

Contact lists can be a bit of effort to create initially but almost certainly will be very useful under a variety of circumstances, including times of crisis.

Ideally, you should keep two. The first would be a detailed list that will be kept in your fireproof safe and the second one, about the size of a credit card, will have essential phone numbers and will reside in your wallet or purse.

For your detailed list, try to include as much information as possible since it may be used by someone else on your behalf. Details like e-mail, phone numbers, addresses, and short comments describing the contact are all very helpful.

The list of people that should be on your detailed contact list should include the following:

- Doctor(s)
- Insurance companies
- Banks that you deal with
- Key family members
- All people who are referenced in your will

For your small contact list, the one that you will keep in your wallet or purse, consider getting it laminated. This can be done for just a few dollars at a local print store and will protect it against wear and water damage. This list can be considered your "ICE" (In Case of Emergency) list and will include the phone numbers and other contact information for one or two key people. In addition to your wallet, you may also want to consider putting this list in a prominent place in your home, like on your fridge or near your home phone.

Emergency contact plan

Do you know how to contact your key friends and relatives if a major crisis occurs? A hurricane, flood or major terrorist attack can knock out telecommunications hardware for weeks or longer.

Make an agreement with your friends and relatives that if something catastrophic were to happen to either you or them, that you will reach each other through sites like the Red Cross "Safe and Well" site. This is super easy to do and can be done with a quick conversation with the people closest to you.

Do you still have a **landline telephone**? If so and you only have cordless phones, **consider buying or keeping an old cord-based phone**. These will work even if the power is out.

Emergency plans for older adults

If you are an older adult or know of one, you may want to consider the following points **in addition** to the ones already mentioned:

- If you live in the United States, consider signing up for SMART911. This will allow you to give information such as medications used and mobility issues that will appear to the 911 operator if you make a 911 call.
- Make sure that a trusted neighbor or landlord has a copy of your ICE (In Case of Emergency) contacts.

Emergency preparedness for children

There are situations that can arise where our children will have to react to some type of emergency. It could be as simple as you not showing up to pick them up from school or something more serious like you having a debilitating injury at home.

Teaching them a few simple things and precautions will only take a few minutes to do but can make a big difference to you and them if needed. Please consider the following:

- Have them (at an age-appropriate time) remember a few key phone numbers, such as your home phone, office phone, and neighbors' number.
- Teach them how and when to use 911 (including how to use a cell phone, pressing the CALL button, and how to use the emergency function).
- Have them take part in your home emergency planning and practice sessions.
- Go over situations with them such as if you do not show up to pick them up from school or if they become lost.

Emergency plans and meeting points

If your smoke detector or carbon monoxide detector goes off, does everyone in your house know what to do? Does everyone know the escape plan from every room in your house? Do they know how to use your fire extinguisher? How about what to do after exiting the house, do you have a meeting point? Where is the safest place if there is a tornado or earthquake?

A fire or other emergency can happen in any home and a few minutes of preparation now regarding the above questions can literally make the difference between life and death.

Do you know how to turn off some or all of your utilities during an emergency? For instance, you may need to turn off your electricity because of a pending flood event. During an evacuation order, officials may tell you to turn off one or more of your utilities before leaving your home. Everything that you learn now, while not under stress, will make dealing with a real emergency that much easier.

Make sure to include your children in your planning and that they understand the importance.

Photos

Do you have important photos? With the digital world today and pictures being captured by cameras that are in every cell phone, you may think your precious memories are safer than the old fashion printed pictures. The truth is exactly the opposite!

While it is true that printed pictures can suffer physical damage, let us look at all the ways your digital pictures can be lost:

- How often do you backup your pictures from your cell phone? If you do not, then if your cell phone breaks or is lost, so are your pictures.
- Do you realize that even if your pictures are backed up on CD's, portable hard drives or USB sticks, the lifetime of these mediums is no more than 20 years and is typically less than 10?
- With the ease of taking pictures, even if you do backup your pictures, can you find the one you are looking for out of the thousands you have?
- In a fire, unless you have a fireproof safe, your digital storage medium will suffer the same fate as a printed picture.

Suggestions for making sure that your precious memories stay safe:

- Store printed or digital copies in a fireproof safe
- Store select precious images to a picture storage site on the internet.
- Scan important printed pictures and store these on a picture storage site as well.
- Share your pictures! Make an electronic copy of important pictures, put them on a flash drive, and give them to your siblings and close friends.

Insurance

With weather and other natural events like forest fires and flooding becoming more unpredictable, you can expect to make more frequent insurance claims. Though you would like to be insured for everything that you are vulnerable to, you need to know there is a good chance that your current policy does not. Knowing what is covered and what is not is important and with this knowledge, you can adjust your policy if you wish to.

Review your policies or talk to your insurance agent to see if the following are covered or not. Typically, they are not:

- Flooding because of sewer backup.
- Flooding in general.
- Freezer content loss due to extended power loss.
- Mold and insect damage.
- Does your policy cover replacement cost or pro-rated cost? For instance, if your fridge is 10 years old and you have a house fire, does your insurance policy cover the price of a new fridge or a 10-year-old one?
- Does your policy cover open-peril or specific-peril clauses?

The general point is for you to understand what is in your policy **before the emergency**. This will lower your stress during an emergency since **you will know** what is covered instead of wondering and worrying if something is covered.

Fireproof Safe

Putting in hard work to get all this paperwork together will do little good if you cannot locate it during a time of crisis or if you have a home fire. A good fireproof safe is nearly an essential investment for every home. Not only will it protect your critical documents during a fire, but it also provides a central place for them, leaving no question as to where they are.

Sustainability issues have increased fire risk. In California alone in 2018, there were over 6,000 forest fires that burned over 1.5 million acres and destroyed thousands of homes. But even without the threat of increased fire risk, it still makes sense to have a fireproof safe. If your home is damaged, this is the time that you most likely will need items that should be stored in the fireproof box.

A starting list of items to keep in your fireproof box include:

- Safe deposit box keys.
- Original social security cards.
- Original birth and marriage certificates.
- Insurance policies.
- Vaccination papers for your family, including your pets.
- Important banking papers including bank accounts, investments, loans, mortgages, and retirement accounts.
- Important legal documents including wills, power of attorney, healthcare proxies, and wills where you are the executor.
- Contact list.
- Electronic copy (flash drive, external hard drive) of photos (note that this should be recopied every 5 years).
- Electronic copy of important papers that you do not need the originals of.
- Vehicle titles.

There are things that you should not keep in your safe. This would include large sums of money. A general rule of thumb is to keep things in your fireproof safe that are valuable to you but not to a thief.

Your Pets

During emergencies, it is easy for a pet to escape or get lost. A simple way to increase the chances of being reunited is to "chip" your pet. When found, this chip can be scanned and cross-referenced back to you and your contact information.

If you have pet health insurance, a quick thing to check is if it covers heatstroke or not.

Basic First Aid/CPR training

As mentioned back in the "Continuous Education" section, first aid training can be useful in so many situations and can be obtained free or at a very low cost.

Phone Apps

Here are a few ideas for apps to put on your smartphone that may come in handy during emergencies:

- First Aid Guide
- Flashlight
- FEMA
- Medical ID
- Disaster Alert
- ICE apps, like QuickICE

Reference books and websites

During an emergency, Alexia and other internet search engines may not be available. Your phone may get lost or broken. Having a physical book or two may make the difference. One suggestion is the SAS Survival Guide by John "Lofty" Wiseman. More details on this book and other suggestions are in the "Further Study" chapter at the end of this section.

Sometimes you have a day or two to prepare for an imminent emergency, like a hurricane or blizzard, at other times you do not. The following is a listing of government sites that can assist you and will include forecasts and evacuation information:

www.ready.gov

A website that is run by FEMA and is packed with information.

www.getprepared.gc.ca

A website run by the government of Canada on emergency preparedness.

Emergency Supplies

What does a cell phone in the toilet, an indoor dog fart, and an ice storm have in common? They are all emergencies and the degree that you are impacted will depend on what you have in hand at the time that they occur. Emergency supplies will supplement what you normally have on hand and will greatly lower the impact an event can have on you and your family.

What is important is getting the right mix of emergency supplies and having them in the right place. **Unfortunately, most of the government emergency websites and even those run by disaster service organizations have outdated, misleading or incorrect information in this area.**

> An example: from the Homeland Security website (https://www.ready.gov/build-a-kit) April 5, 2019
>
> "To assemble your kit, store items in airtight plastic bags and put your entire disaster supplies kit in one or two easy-to-carry containers such as plastic bins or a duffel bag."

This doesn't sound like a bad idea except that the first item on their list is water, 3 gallons (11 liters) per person or 25 lbs (11.4 kg). Logistically, this means that your "easy-to-carry" containers for a family of four will weigh around 150 pounds (68 kg) once the rest of the supplies are added. Not too practical if you had to evacuate quickly. A more sensible plan is needed.

For the purposes of this section, we will classify an emergency as some event, either natural or accidental, that puts you or your family at risk. It does not include the risk of someone intentionally trying to do you harm. Some examples include:

- Weather-related events such as blizzards, flooding or tornadoes.
- Water, electricity, natural gas or some other utility disruption for an extended period.

- Physical injury due to an accident.
- Communication outage that prevents critical communication between you and your family.

It is simply not practical to carry with you all the supplies needed all the time to handle any emergency. This means we need to sort out what items we need and exactly where they should be stored.

Your Home

If you are in or near your home when a crisis strikes, you have two choices of what to do. Either stay and ride the emergency out or evacuate.

The decision to stay usually means that there is no immediate threat to life and is usually the result of a power outage caused by an ice storm or hurricane, or a flooding event on nearby roads that makes leaving impossible. Regardless of the reason, there is the potential that you may need to survive for a week or longer on your own before the crisis passes and without the aid from disaster services.

The decision to voluntarily evacuate or perhaps be forced out by mandatory evacuation means that there is an imminent danger to you and your family. By evacuating, you are going somewhere safer and while this may be a stressful and dangerous event, the time you need to get to this destination is short, usually less than a day and rarely over a few.

These two different scenarios, leaving or staying, require many of the same emergency supplies but there are differences and for this reason, I would recommend that you keep two sets of supplies that complement each other.

The first would be very portable, for instance, in a backpack, and would contain everything that your family would need for a few days when away from your home. The second would stay in your home and could be stored in your basement or a closet. This second set of supplies would be used in conjunction with your portable supplies for when you decide to stay in your home during the crisis.

The table below contains a generalized list for both kits. You can start with this and then customize each to meet your specific needs. Most, if not all items can be purchased online or in either a camping or hardware store.

If you wish, you can purchase "ready-to-go" 72-hour portable kits through online retailers or through organizations like the Red Cross. If you do decide to purchase one of these ready-to-go kits, please remember to cross-reference the contents to the list below and add any items, like copies of your important papers, that may be missing.

The one item from the table below that is rather bulky is the water needed for the home kit. For a family of four, having a supply of about 50 gallons (189 liters) is recommended (enough to last about 2 weeks). An alternative for at least some of this, if you have one, is by draining water from your hot water heater (remember to turn the heater off first and to filter the water through something like a coffee filter before drinking).

Another important point is to stock your home kit with "ready-to-eat" food. You may think this is not important since you always have lots of food in your home, but will you? Remember that in a time of crisis, if your electricity is off, the food in your refrigerator and freezer will be spoiled within a few days.

You may have already experienced this or at least seen it on television. A hurricane is predicted to hit an area in a few days and people line up at stores to get supplies, usually to be greeted by empty shelves. Stocking this food now is easy and stress-free.

Finally, and perhaps the most important point:

Do not put anything in your kit that you do not know how to use.

Item	Home Kit	72-Hour Portable Kit
Water	12 gallons (45 liters) per person.	½ gallon (2 liters) per person in small pouches.
Water purification tablets, straw, or filter	Enough for 20 gallons (75 liters) per person.	Enough for 5 gallons (20 liters) per person.
Food	Enough long shelf life food to last 2 weeks.	Energy bars and snacks, at least 4,500 calories per person.
First aid kit	Well-stocked kit.	Smaller portable kit
Hand crank combo radio, flashlight, phone charger	Get from 72-hour kit.	Get a good quality one. Radio should pick up NOAA weather channels.
Dust masks	Get from 72-hour kit.	Re-usable N95 per person. Ensure children have proper-sized masks.
Duct tape	Get from 72-hour kit.	If the space station needs this, so do you!
Whistle	Get from 72-hour kit.	Used to get attention, if necessary. One per person.
Sanitation items	Enough to last for 2 weeks.	Enough for each person for a few days.
Multi-tool	Get from 72-hour kit.	Get a good one that includes at least a knife, plier, can opener, and screwdriver.
Copy of critical papers	Get from 72-hour kit.	Driver's license, passport, contact list, deed to property.
Gloves	Get from 72-hour kit.	1 set of latex or other medical glove, 1 set of heavy-duty work gloves.
Firestarter and kindling	A good supply of waterproof matches, you can also use striker from 72-hour kit	Ideally, a magnesium striker, but waterproof matches or lighter are okay.

Kindling for fire	If you have a wood stove, fireplace or fire pit and have wood, then you can make your own kindling.	A dozen or so cotton balls, pre-soaked in petroleum jelly.
Portable cooking stove	A portable camping stove with enough fuel for 2 weeks. Cannot rely on barbeque since it may run out of fuel. Make sure in-door safe or use outdoors to avoid carbon monoxide poisoning.	This is not a camping trip! Though several 72-hour kits include a camping stove, I do not see the need for it. You can survive on cold food for a day or two. There are exceptions, such as if you need to heat formula for an infant.
Cash	Get from 72-hour kit	Small-denomination bills and coins. Ideally enough to buy a meal or two.
Entertainment	Get from 72-hour kit or find something within your home.	Books, cards, etc. Nothing electronic unless you can charge it manually.
Maps of local area	Get from 72-hour kit.	Good printed map of at least a 100-mile radius.
Extra set of house and car keys	Get from 72-hour kit.	If keys have a battery, store in a plastic bag.
Flashlights	USB rechargeable LED flashlight. One per person.	One USB rechargeable LED flashlight. Can be charged by your hand crank charger.
Candles	Supply to last 2 weeks.	A 50-hour candle or a couple of long-lasting light sticks.
Rope	Get from 72-hour kit.	50 ft (15 m) or more of nylon cord or rope.
Blanket / Sleeping bags	Get from home or from 72-hour kit.	Emergency blanket or blanket/sleeping bag

		combination – one per person.
Clothes	Get from home or from 72-hour kit	One change of clothes
Other items to consider depending on location and your specific situation		
Pet food and medicine	Enough for 2 weeks.	Enough for a few days.
Prescription medication, supplies, spare glasses, hearing aid batteries.	Enough for 2 weeks. Watch shelf life and rotate as necessary.	Enough for a few days. Watch shelf life and rotate as necessary.
Baby supplies – Diapers, formula, food.	Enough for 2 weeks.	Enough for a few days.
Rain ponchos	Get from 72-hour kit.	One for each person.
Extra socks	Get from home or 72-hour kit.	One pair per person. Important to keep feet as dry as possible.
Hats or toques.	Get from home or 72-hour kit	One per person.
Bug Spray	Get from home or 72-hour kit	Enough for a few days.

Your Vehicle

Though you may encounter a crisis while in or near your vehicle, the supplies that you should have in it varies considerably depending on the environment that you travel in. The following list shows common items to consider. Apply common sense when building this kit. If you live in an area that is warm year-round, you will not need a snow shovel. If you want to save a bit of time, you can buy pre-made kits.

The one *gotcha* with vehicle emergency kits is water. Though you can store water for years. If you live in an area that has freezing winter conditions, that bottle of water will turn into a block of ice which you may not be able to thaw out very easily if your car breaks down and you are stranded. Those of you that live in areas of extreme heat also must be careful, as the heat, over an extended period, will degrade plastic water bottles and contaminate the contents.

Given these issues, the practical advice is to take some water with you if you are intending on making a trip in adverse conditions and on roads that have little traffic where you may get stranded for an extended period.

One final point. If your car gets stuck in snow, make sure that the exterior exhaust pipe is clear from snow to prevent carbon monoxide buildup inside your vehicle.

Vehicle Emergency Kit Suggestions	
Jumper cables	First aid kit
Sturdy ice scraper and brush	Collapsible shovel
Work gloves	Rock salt, sand or kitty litter
Warm blanket	Extra windshield washer fluid
Orange construction marking tape. Used for drawing attention to your vehicle	Non-perishable snacks
50-hour candle	Waterproof matches
Hand crank radio with USB charging port	Poncho
Flares or reflective triangle	Reflective traffic safety vest
Multi-tool	Flare
Whistle	Toilet paper
Safety hammer with seat belt cutter. **Keep this item accessible, like in the glove box**.	

On your person

When you are away from your home and vehicle, you still can encounter a crisis. While it is not practical to carry around supplies, there are three things that nearly all of us have on them that we can stock up and will be useful.

On-Person Emergency Kit	
Your Cell Phone	There are several apps that can help in times of need. First Aid Manual, Flashlight, FEMA alerts, and ICE (In Case of Emergency) contact information are a few examples.
Your Wallet	What happens if your phone runs out of power, breaks or gets lost and you need to get in touch with a family member? Do you have their numbers memorized? Having key phone numbers on a slip of paper in your wallet will handle this emergency.
Your Brain	Take a first aid course. Online and classroom versions are available. They are cheap or free and only take a few hours.

Tips on Water

You probably have heard the expression that you can only live 3 minutes without air, 3 days without water, and 30 days without food. This is directionally correct, but regarding water, 3 days may be the optimal case. If your emergency is during a heatwave, lack of water even for a few hours can cause you to start dehydrating and this is especially true for babies and children.

During an emergency, it is also important to make sure that whatever water you do have access to is safe to drink, since the last thing you want to do is get sick when mobility is limited and medical facilities may be overflowing. When in doubt, do not drink tap or other sources of water without boiling first.

Water purification tablets

Water purification tablets work to kill many viruses and bacteria that can be in untreated water. The chemical make-up varies but the common chemicals in tablets are chlorine, iodine, and chlorine dioxide.

Most of the tablets are not meant for long-term use and there are still some bacteria and viruses that are not killed by the tablets, so these should only be used as a last resort.

Alternatives to tablets

There are alternatives to tablets such as water purification straws and filters. In general, these will run the same risk as tablets and should only be used as a last resort.

One interesting source of water during an emergency is your hot water heater. Before using it, ensure that the unit is turned off. Since the drain on your water tank is on the bottom, there is a chance that water from it may have sediment in it. Filtering it through a coffee filter or clean towel should be all that is needed to have clean, drinkable, water.

Can water expire?

I always find it amusing to see an expiration date on bottled water but there is some validity to it. Though water itself does not "expire", the plastic bottle that it is usually stored in does, especially if it has been exposed to heat or sunlight for an extended period.

If you have just opened a bottle of water, how long is it good for? The answer to this question depends on a bunch of factors, but if you want to play it very safe, you should consume that open bottle within 6 hours. If you store an opened bottle in the fridge, it should remain quite safe for at least 5 days (198).

Random Notes

The following is a collection of suggestions and information that did not quite fit into the other sections that you may want to consider when planning for an emergency:

- If your power goes out, a full freezer will keep foods frozen for about 2 days. A half-full one, about a day.
- Kind of a no-brainer, but if it is winter outside, put your expensive frozen food outside!
- You may own one or more flashlights, but do you know exactly where they are? Always keep one in the same location so you can find it easily in the dark.
- If you do make a last trip to the grocery store, consider that you may lose power. Buy substitutes, like milk powder and dried fruit.
- Do not forget to store food, water, and medicine for your pet(s).

Further Study – Impact and Readiness

The following list of books and websites can be used to get more detailed information on the impacts of the trends along with ways to prepare and adapt. They are "kind of" in the order presented in this section, but not exactly, so scan through the entire list if you are looking for a specific topic. Children's books are grouped together near the end of the list.

Cover/Logo	Age	Title/Description/Link
	Mental Health Impacts	
	10+	Good Pictures, Bad Pictures: Porn Proofing Today's Young Kids -Kristen A. Jensen, Debbie Fox
	14+	The Anxiety Workbook for Teens: Activities to Help You Deal with Anxiety and Worry -Lisa M. Schab LCSW
	14+	The Relaxation and Stress Reduction Workbook for Teens: CBT Skills to Help You Deal with Worry and Anxiety (Instant Help) -Michael A. Tompkins PhD ABPP, Jonathan R. Barkin PsyD

	16+	Screenwise: Helping Kids Thrive (and Survive) in Their Digital World. - Devorah Heitner, PhD
	16+	The Art of Screen Time – How Your Family Can Balance Digital Media and Real Life -Anya Kamenetz
	16+	Emotional Recovery from Situation Anxiety – How You Can Feel Safe Again –Richard Schwindt
	16+	Sleep Smarter: 21 Essential Strategies to Sleep Your Way to A Better Body, Better Health, and Bigger Success Hardcover – March 15, 2016 –Shawn Stevenson

	16+	The Anxiety & Worry Workbook –David A. Clark, Aaron T. Beck
Impact on Physical Health		
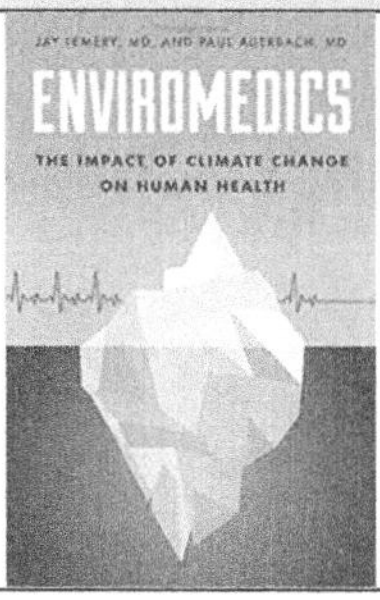	16+	Enviromedics: The Impact of Climate Change on Human Health –Jay Lemery, Paul Auerbach
	16+	Prescription for Nutritional Healing, Fifth Edition: A Practical A-to-Z Reference to Drug-Free Remedies Using Vitamins, Minerals, Herbs & Food Supplements –Phyllis A. Balch CNC
	16+	Kiss the Ground: How the Food You Eat Can Reverse Climate Change, Heal Your Body & Ultimately Save Our World –Josh Tickell

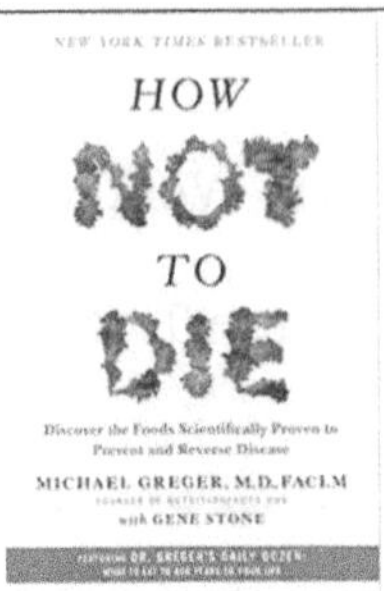 	16+	How Not to Die – Discover the Foods Scientifically Proven to Prevent and Reverse Disease –Michael Greger M.D. FALCM. Gene Stone
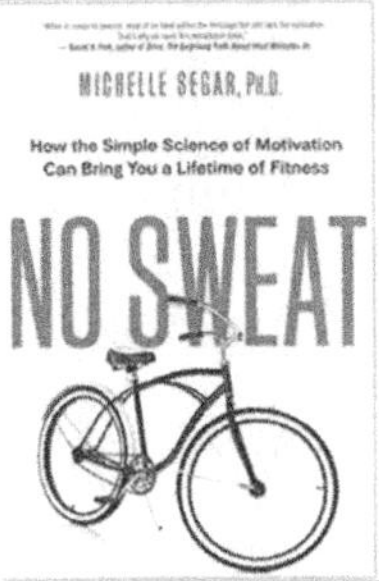 	16+	No Sweat: How the Simple Science of Motivation Can Bring You a Lifetime of Fitness –Michelle Segar
	Employment and Lifestyle	
 	16+	Stretch: How to Future-Proof Yourself for Tomorrow's Workplace – Karie Willyerd
 	16+	Buyology: Truth and Lies About Why We Buy – Martin Lindstrom and Paco Underhill

	16+	Why We Buy: The Science of Shopping – Martin Lindstrom and Paco Underhill
	16+	The High Price of Materialism – Tim Kasser
	16+	The Minimalist Home: A Room-by-Room Guide to a Decluttered, Refocused Life – Joshua Becker You may also be interested in this book, also written by Joshua Becker. The More of Less: Finding the Life You Want Under Everything You Own
	16+	The Life-Changing Magic of Tidying Up: The Japanese Art of Decluttering and Organizing – Marie Kondō

	16+	Finding Your Way in a Wild New World: Reclaim Your True Nature to Create the Life You Want – Martha Beck You may also be interested in this book, also written by Martha Beck. Finding Your Own North Star: Claiming the Life You Were Meant to Live
	16+	The Robots are Coming: A Human's Survival Guide to Profiting in the Age of Automation – John Pugliano
	16+	The Traveler's Gift – Seven Decisions that Determine Personal Success – Andy Andrews
	16+	How to Stop Living Paycheck to Paycheck – A proven path to money mastery in only 15 minutes per week – Avery Breyer

	14+	Feasible Planet – A Guide to More Sustainable Living – Ken Kroes
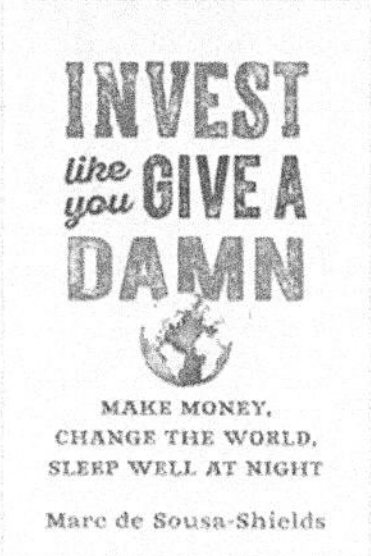	16+	Invest Like You Give a Damn – Make Money, Change the World, Sleep Well at Night – Marc de Sousa-Shields
	16+	Invest for Good - A Healthier World and a Wealthier You – Mark Mobius, Carlos von Hardenberg, Greg Konieczny
	16+	Biblically Responsible Investing - For God's Glory and Your Joy – Robert Netzly
	16+	Creation Tithing – A Primer – Anne Ipsen, Ken Kroes

<table>
<tr><th colspan="3">Social Impact</th></tr>
<tr><td>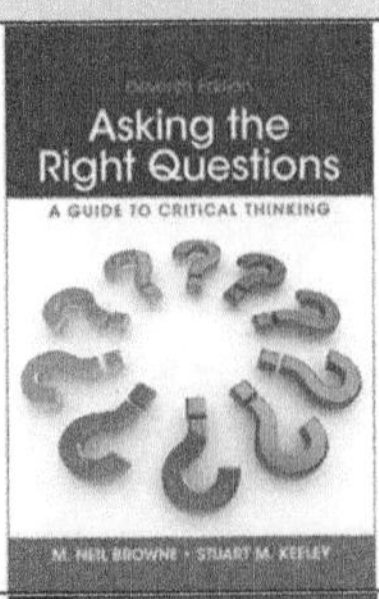
</td><td>16+</td><td>Asking the Right Questions – A Guide to Critical Thinking

–M. Neil Brown, Stuart M. Keely</td></tr>
<tr><td>
</td><td>16+</td><td>How We Know What Isn't So – The Fallibility of Human Reason in Everyday Life

–Thomas Gilovich</td></tr>
<tr><td>
</td><td>16+</td><td>Thinking, Fast and Slow

–Daniel Kahneman</td></tr>
<tr><td>
</td><td>16+</td><td>The World's Religions

–Huston Smith</td></tr>
</table>

	16+	Religious Literacy – What Every American Needs to Know – and Doesn't – Stephen Prothero
	16+	The Madhouse Effect: How Climate Change Denial Is Threatening Our Planet, Destroying Our Politics, and Driving Us Crazy – Michael Mann and Todd Toles
	16+	Kid Confidence: Help Your Child Make Friends, Build Resilience, and Develop Real Self-Esteem – Eileen Kennedy PhD
	16+	The Parent's Guide to Climate Revolution: 100 Ways to Build a Fossil-Free Future, Raise Empowered Kids, and Still Get a Good Night's Sleep – Mary DeMocker, Bill McKibben

Emergency Readiness		
	16+	SAS Survival Guide: The Ultimate Guide to Surviving Anywhere – John "Lofty" Wiseman.

Web Sites related to Family Online Use Plan
Healthy Children A comprehensive check list for youth online usage and a tool to build up a personalized plan. www.HealthyChildren.org/MediaUsePlan
American Academy of Pediatrics Several resources for parents. www.aap.org/en-us/advocacy-and-policy/aap-health-initiatives/Pages/Media-and-Children.aspx

Good Informational Web Sites
Grist A very good site with news and advice on mostly environmental trends. www.grist.org
The Guardian Good, unbiased reporting of news on many of the trends brought up in this book. www.theguardian.com/international
The Revelator In depth articles on a wide range of topics. www.therevelator.org

Children's Books, Websites, Games, Toys, Magazines, and Papers		
	10-14	Climate Change – Discover How It Impacts Spaceship Earth –Joshua Sneidman, Erin Twamley, Mike Crosier
	6-12	What to Do When You Worry Too Much: A Kid's Guide to Overcoming Anxiety (What-to-Do Guides for Kids) -Dawn Huebner
	7 – 11	The Tantrum that Saved the World – Megan Herbert & Michael Mann
	10+	The Fallacy Detective: Thirty-Eight Lessons on How to Recognize Bad Reasoning – Nathaniel Bluedorn, Hans Bluedorn, Rob Corley, Tim Hodge
	4+	Skill Sharpeners Critical Thinking There are several books in this series, from pre-kindergarten through grade 6 – Evan-Moor Corp

	7 – 11	What Every Child Should Know About Climate Change – Baby Professor
	6-9	The Lorax (Classic Seuss) – Dr. Seuss
	10+	It's Your World: Get Informed, Get Inspired & Get Going! – Chelsea Clinton
		Amber's Magical Savings Box: First Interactive Lesson on Earning and Saving Money! – Rachael Hanibel
	5+	Iggy Peck, Architect – Andrea Beaty, David Roberts If you like this book, there is a whole series.
	8-11	Unofficial Minecraft STEM Lab for Kids: Family-Friendly Projects for Exploring Concepts in Science, Technology, Engineering, and Math – John Miller & Chris Fornell Scott

		You may also want to check out other Minecraft related software and books related to teaching STEM.
	*9+*	A Wild Child's Guide to Endangered Animals – Millie Marotta
	*8-12*	National Geographic Animal Encyclopedia: 2,500 Animals with Photos, Maps, and More! – Lucy Spelman
	*3 - 7*	Chirp Magazine– Stories, puzzles, and age-appropriate activities. Search for "Chirp Magazine" with an internet browser to preview, order a subscription (10 per year), or buy old issues.
	*6-14*	Nat Geo Kids Magazine – Focuses on animals, science, nature, technology, exploration, adventure, and more. Search for "Nat Geo Kids" with an internet browser to preview or order a subscription (available in print and electronic format).
	*3-6*	Nat Geo Little Kids Magazine – Dazzling pictures, stories, and picture games. Search for "Nat Geo Little Kids" with an internet browser to preview or order a subscription (available in print and electronic format).

	*2-6*	High Five Magazine – Stories, puzzles, and age-appropriate activities. Search for "High Five Magazine" with an internet browser to preview or order a subscription (available in print and electronic format, 12 issues per year).

YouTube Channels

A great selection of videos, podcasts, and other content for youth are available as YouTube Channels. Here is a starting list that can be accessed by going to www.youtube.com and entering the names listed below. As with anything on the internet, content should be monitored to ensure that it is age appropriate.

Smithsonian Channel
PBS Kids
Ted Ed
Crash Course Kids
Steve Spangler's Sick Science
SoulPancake

Nasa Kids Club

Student games on NASA websites entertain as well as educate. They support national education standards in STEM for students in kindergarten through fourth grade and have a great deal of information on the natural trends brought up in this book.

www.nasa.gov/kidsclub/index.html

Admongo

Created by the Federal Trade Commission and PR agency Fleishman-Hillard, this site helps youth understand how advertising is done and how it can influence them. Unfortunately, the interactive game is no longer available, but there is still a text version with great information.

www.admongo.gov

Taking it Global (TIG) One of the world's leading networks of young people learning about, engaging with, and working towards tackling global challenges. www.tigweb.org
STEM Toys There are so many out there and each child is unique, so I would recommend doing an internet search for "Best STEM Toys" or checking out the two links below which have a bunch of good recommendations. www.wired.com/gallery/best-stem-toys-for-kids/ www.heavy.com/toys/2018/09/best-stem-toys/
STEM Games There are a bunch of STEM board games that people of all ages can enjoy. Examples include, Bioxels (build your own video game), Mastermind, Code and Go Robot Mouse, Safe Breaker, and Qbitz. More can be found by an internet search for "STEM Board Games". There are also many STEM-type games for your tablet, PlayStation, Nintendo or other gaming systems. Some examples include Nintendo Labo, Frog Dissection, and Math vs. Zombies. Try a search for your specific system.
Code.org Run by a non-profit organization that is dedicated to expanding access to computer science in schools. Free online courses and projects.
An excellent and in-depth paper titled "The Contribution of Early Childhood Education to a Sustainable Society" edited by Ingrid Pramling Samuelsson and Yoshie Kaga. Do an internet search for this PDF paper.
"Education for Sustainable Development in the Early Years" by John Siraj-Blatchford, Kimberly Caroline Smith, and Ingrid Pramling Samuelsson. This document should be readily available through an internet search of the title.

<table>
<tr><th colspan="3">Books on Prepping</th></tr>
<tr><td>
</td><td>16+</td><td>Prepper's Long-Term Survival Guide: Food, Shelter, Security, Off-the-Grid Power and More Life-Saving Strategies for Self-Sufficient Living (Preppers)

– Jim Cobb</td></tr>
<tr><td>
</td><td>16+</td><td>Survival Medicine & First Aid: The Leading Prepper's Guide to Survive Medical Emergencies in Tough Survival Situations

– Beau Griffin</td></tr>
</table>

Conclusion

On a planetary scale, I do not see the global motivation or actions needed to change the course of the trends over the next several decades. Yes, this is depressing, but it is critical to not let this push your ecological anxiety to a level where you start to believe that there is nothing you can do.

Now is the time to decide how far you are willing to go to set yourself up for success and to reduce the impact the trends will have on you and your family. While some of the tips in this book are easy and quick to do, the ones with the biggest benefit, like moving, learning new skills, expanding your social currency, and mentoring youth take commitment.

The trends will not wait for you to be ready. Pick a few relevant suggestions from this book and act upon them, then pick a few more. Select one or two of the many recommended books to get better informed. Get your family involved, especially the youth.

Our population is peaking because we are running into boundaries like the ones mentioned in this book. It is easy to become afraid of the future but with an understanding of the true situation along with the impacts, you can decrease your level of ecological anxiety and adapt to prosper in the coming years.

From the Author

Thank you for reading Feasible Living. If you enjoyed this book (or even if you didn't), please visit the site where you purchased it and write a brief review. Your feedback is important to me and will help other readers decide whether to read the book, too.

My website is www.feasibleplanet.com and I can be contacted at ken@feasibleplanet.com.

References

1. [Online] [Cited: 01 29, 2019.] https://www.culligannation.com/is-chlorinated-water-safe-to-drink/.

2. [Online] [Cited: 08 18, 2018.] https://www.ewg.org/research/report-110-million-americans-could-have-pfas-contaminated-drinking-water.

3. [Online] [Cited: 07 23, 2018.] https://www.womenshealthmag.com/health/a19566224/is-tap-water-safe/.

4. [Online] [Cited: 08 18, 2018.] https://www.theguardian.com/environment/2017/sep/06/plastic-fibres-found-tap-water-around-world-study-reveals.

5. [Online] [Cited: 08 18, 2018.] https://www.theguardian.com/environment/2018/mar/15/microplastics-found-in-more-than-90-of-bottled-water-study-says.

6. [Online] [Cited: 01 30, 2019.] https://www.usatoday.com/story/news/nation/2014/06/25/beaches-polluted-pollution-water-runoff-sewage-bacteria/11349409/.

7. [Online] [Cited: 18 19, 2018.] http://www.who.int/news-room/fact-sheets/detail/drinking-water.

8. [Online] [Cited: 08 20, 2018.] https://www.theguardian.com/environment/2016/feb/12/four-billion-people-face-severe-water-scarcity-new-research-finds.

9. [Online] [Cited: 08 20, 2018.] https://borgenproject.org/sustainable-agriculture-in-yemen/.

10. [Online] [Cited: 08 20, 2018.] https://www.circleofblue.org/2015/world/groundwater-depletion-stresses-majority-of-worlds-largest-aquifers/.

11. [Online] [Cited: 01 30, 2019.] https://therevelator.org/shrinking-groundwater/.

12. [Online] [Cited: 05 03, 2018.] https://www.nationalgeographic.com/environment/global-warming/big-thaw/.

13. [Online] [Cited: May 09, 2018.] https://www.scientificamerican.com/article/shrinking-mountain-glaciers-are-affecting-people-downstream/.

14. [Online] [Cited: 08 20, 2018.] https://savenaturesavelife.wordpress.com/2010/03/27/major-rivers-that-do-not-reach-the-sea-anymore-share/.

15. [Online] [Cited: 08 20, 2018.] https://www.bbc.com/news/world-42982959.

16. [Online] [Cited: 08 20, 2018.] https://water.usgs.gov/edu/wateruse-trends.html.

17. [Online] [Cited: 04 29, 2019.] https://www.climate.gov/news-features/understanding-climate/climate-change-ocean-heat-content.

18. [Online] [Cited: 01 30, 2019.] https://phys.org/news/2018-03-north-atlantic-salty-blame-climate.html.

19. [Online] [Cited: 01 30, 2019.] https://theenergymix.com/2017/03/28/mann-study-finds-atlantic-jet-stream-slowing-down-as-arctic-warms/.

20. [Online] [Cited: 08 20, 2018.] https://climate.nasa.gov/news/2680/new-study-finds-sea-level-rise-accelerating/.

21. [Online] [Cited: 02 23, 2019.] https://www.huffingtonpost.ca/2018/08/25/climate-change-home-values_a_23509303/.

22. [Online] [Cited: 04 30, 2019.] https://www.ucsusa.org/global-warming/global-warming-impacts/co2-ocean-acidification.

23. [Online] [Cited: 04 30, 2019.] https://phys.org/news/2018-07-ocean-acidification-major-impact-marine.html.

24. [Online] [Cited: 04 30, 2019.] https://www.pmel.noaa.gov/co2/story/Ocean+Acidification's+impact+on+oysters+and+other+shellfish.

25. [Online] [Cited: 08 06, 2018.] http://www.who.int/airpollution/.

26. [Online] [Cited: 04 30, 2019.] https://e360.yale.edu/features/redrawing-the-map-how-the-worlds-climate-zones-are-shifting.

27. [Online] [Cited: 04 30, 2019.] https://climate.nasa.gov/news/2436/co2-is-making-earth-greenerfor-now/.

28. [Online] [Cited: 04 24, 2019.] https://airqualitynews.com/2018/08/08/ozone-pollution-harming-crop-yields-global-study-suggests/.

29. [Online] [Cited: 01 31, 2019.] https://grist.org/food/theres-arsenic-in-your-rice-and-heres-how-it-got-there/.

30. [Online] [Cited: 09 05, 2018.] http://www.fao.org/news/story/en/item/1126971/icode/.

31. [Online] [Cited: 09 05, 2018.] https://www.igb-berlin.de/en/news/underestimated-threat-land-based-pollution-microplastics.

32. [Online] [Cited: 09 05, 2018.] https://www.unenvironment.org/news-and-stories/story/plastic-planet-how-tiny-plastic-particles-are-polluting-our-soil.

33. [Online] [Cited: 02 01, 2019.] https://www.globalagriculture.org/report-topics/soil-fertility-and-erosion.html.

34. [Online] [Cited: 02 01, 2019.] https://www.popsci.com/topsoil-agriculture-food#page-3.

35. [Online] [Cited: 01 31, 2019.] https://www.washingtonpost.com/news/wonk/wp/2013/11/09/no-till-farming-is-on-the-rise-thats-actually-a-big-deal/?utm_term=.a22b156956dd.

36. [Online] [Cited: 01 31, 2019.] https://iopscience.iop.org/article/10.1088/1748-9326/10/2/024019.

37. [Online] [Cited: 09 07, 2018.] https://www.theguardian.com/environment/ng-interactive/2018/jun/27/one-football-pitch-of-forest-lost-every-second-in-2017-data-reveals.

38. [Online] [Cited: 09 08, 2018.] https://www.theguardian.com/global-development-professionals-network/2017/jan/23/destroying-rainforests-quickly-gone-100-years-deforestation.

39. [Online] [Cited: 02 01, 2019.] http://www.borealforest.org/world/world_overview.htm.

40. [Online] [Cited: 09 08, 2019.] https://www.cbc.ca/news/technology/boreal-forest-being-driven-to-tipping-point-by-climate-change-study-finds-1.3198892.

41. [Online] [Cited: 09 08, 2019.] https://www.nationalobserver.com/2018/04/20/news/worst-case-scenario-trees-big-parts-canadas-boreal-forest-will-probably-die-says.

42. [Online] [Cited: 01 15, 2019.] http://theconversation.com/meet-n2o-the-greenhouse-gas-300-times-worse-than-co2-35204.

43. [Online] [Cited: 01 15, 2019.] https://www.epa.gov/ghgemissions/overview-greenhouse-gases.

44. [Online] [Cited: 01 15, 2019.] https://www.ncbi.nlm.nih.gov/pmc/articles/PMC1247398/.

45. [Online] [Cited: 01 16, 2019.] https://e360.yale.edu/features/can-the-world-find-solutions-to-the-nitrogen-pollution-crisis.

46. [Online] [Cited: 09 09, 2018.] https://minerals.usgs.gov/minerals/pubs/commodity/phosphate_rock/mcs-2018-phosp.pdf.

47. [Online] [Cited: 01 16, 2019.] http://www.bbc.com/future/story/20140314-the-worlds-scarcest-material.

48. [Online] [Cited: 05 06, 2019.] https://en.wikipedia.org/wiki/Extinction_event#Effects_and_recovery.

49. [Online] [Cited: 01 19, 2019.] https://news.mongabay.com/2007/03/70-of-new-drugs-come-from-mother-nature/.

50. [Online] [Cited: 01 19, 2019.] https://www-tc.pbs.org/wgbh/nova/julian/media/lrk-disp-plantmedicines.pdf.

51. [Online] [Cited: 01 19, 2019.] https://www.theguardian.com/environment/2018/jul/05/red-list-research-finds-26000-species-under-extinction-threat.

52. [Online] [Cited: 01 19, 2019.] http://livingplanetindex.org/home/index.

53. [Online] [Cited: 01 19, 2019.] http://livingplanetindex.org/projects?main_page_project=LivingPlanetReport&home_flag=1.

54. [Online] [Cited: 05 06, 2019.] https://www.nationalgeographic.com/environment/2019/05/ipbes-un-biodiversity-report-warns-one-million-species-at-risk/.

55. [Online] [Cited: 02 04, 2019.] https://en.wikipedia.org/wiki/Stem_rust#History_of_stem_rust.

56. [Online] [Cited: 02 04, 2019.] https://agfax.com/2018/09/28/rice-blast-the-most-devastating-rice-disease-in-the-world/.

57. [Online] [Cited: 02 04, 2019.] https://www.theguardian.com/environment/2018/dec/07/deadly-pig-virus-african-swine-fever-us-china.

58. [Online] [Cited: 02 04, 2019.] 003. https://news.sky.com/story/dire-predictions-for-future-of-world-wildlife-11303501.

59. [Online] [Cited: 02 04, 2019.] https://thehill.com/opinion/energy-environment/409002-the-global-climate-refugee-crisis-has-already-begun.

60. [Online] [Cited: 02 04, 2019.] https://www.huffingtonpost.ca/2018/10/25/earth-doesnt-have-enough-fruits-veggies-to-keep-everyone-healthy-study_a_23571970/?utm_hp_ref=ca-homepage.

61. [Online] [Cited: 12 26, 2018.] https://stats.oecd.org/Index.aspx?DataSetCode=POP_PROJ#.

62. [Online] [Cited: 02 16, 2019.] http://www3.weforum.org/docs/WEF_Future_of_Jobs.pdf.

63. [Online] [Cited: 01 28, 2019.] https://driving.ca/general-motors/auto-news/news/gm-cars-automatic-braking-proven-to-reduce-collisions.

64. [Online] [Cited: 11 01, 2018.] https://www.cnbc.com/2017/07/19/half-of-american-jobs-are-at-risk-from-automation-new-study-suggests.html.

65. [Online] [Cited: 11 01, 2018.] https://www.washingtonpost.com/news/wonk/wp/2017/11/30/robots-could-soon-replace-nearly-a-third-of-the-u-s-workforce/.

66. [Online] [Cited: 11 03, 2018.] https://www.stlouisfed.org/publications/regional-economist/fourth-quarter-2017/evolution-american-workplace.

67. [Online] [Cited: 04 24, 2019.] https://www.floridatrend.com/article/24033/strawberry-boss.

68. [Online] [Cited: 01 27, 2019.] https://whatis.techtarget.com/definition/gig-economy.

69. [Online] [Cited: 01 05, 2019.] https://www.washingtonpost.com/politics/2018/12/30/year-unprecedented-deception-trump-averaged-false-claims-day/?utm_term=.a23ebe8548e1.

70. [Online] [Cited: 01 05, 2019.] https://www.pbs.org/wgbh/frontline/article/climate-change-skeptic-group-seeks-to-influence-200000-teachers/.

71. [Online] [Cited: 01 05, 2019.] https://www.ogilvy.com/feed/5-types-of-fake-news-and-why-they-matter/.

72. [Online] [Cited: 01 05, 2019.] https://www.snopes.com/fact-check/did-cdc-flu-shot-causing-outbreak/.

73. [Online] [Cited: 01 05, 2019.] https://globalnews.ca/news/4396939/measles-vaccination-fake-news/.

74. [Online] [Cited: 01 05, 2019.] https://www.thedailybeast.com/the-fake-news-epidemic-in-health.

75. [Online] [Cited: 01 05, 2019.] http://news.mit.edu/2018/study-twitter-false-news-travels-faster-true-stories-0308.

76. [Online] [Cited: 01 05, 2019.] https://www.washingtonpost.com/opinions/fake-news-is-about-to-get-so-much-more-dangerous/2018/09/06/3d7e4194-a1a6-11e8-83d2-70203b8d7b44_story.html.

77. [Online] [Cited: 01 05, 2019.] https://www.buzzfeednews.com/article/davidmack/obama-fake-news-jordan-peele-psa-video-buzzfeed.

78. [Online] [Cited: 01 15, 2019.] http://www.newser.com/story/225645/average-size-of-us-homes-decade-by-decade.html.

79. [Online] [Cited: 05 11, 2019.] https://www.money-zine.com/financial-planning/debt-consolidation/consumer-debt-statistics/.

80. [Online] [Cited: 01 15, 2019.] https://www.psychologytoday.com/us/blog/anxiety-files/201610/the-decline-trust.

81. [Online] [Cited: 05 11, 2019.] https://news.gallup.com/poll/195542/americans-trust-mass-media-sinks-new-low.aspx.

82. [Online] [Cited: 06 05, 2019.] https://www.cnn.com/2012/06/25/tech/web/mcafee-teen-online-survey/index.html.

83. [Online] [Cited: 06 07, 2019.] https://www.marketwatch.com/story/teens-spend-all-their-free-time-on-social-media-and-many-dont-read-anymore-2018-08-20.

84. [Online] [Cited: 06 07, 2019.] https://hackernoon.com/how-much-time-do-people-spend-on-their-mobile-phones-in-2017-e5f90a0b10a6.

85. [Online] [Cited: 02 04, 2019.] https://www.cnbc.com/2017/11/14/richest-1-percent-now-own-half-the-worlds-wealth.html.

86. [Online] [Cited: 02 04, 2019.] http://www.pewresearch.org/fact-tank/2018/09/06/the-american-middle-class-is-stable-in-size-but-losing-ground-financially-to-upper-income-families/.

87. [Online] [Cited: 04 10, 2019.] https://www.forbes.com/sites/panosmourdoukoutas/2018/11/24/debt-not-trade-war-is-chinas-biggest-problem/#5328079e4c4d.

88. [Online] [Cited: 05 14, 2019.] https://www.npr.org/2019/02/13/694199256/u-s-national-debt-hits-22-trillion-a-new-record-thats-predicted-to-fall.

89. [Online] [Cited: 01 21, 2019.] https://www.thebalance.com/current-u-s-federal-budget-deficit-3305783.

90. [Online] [Cited: 01 21, 2019.] https://www.cbsnews.com/news/the-skyrocketing-interest-payments-on-u-s-debt/.

91. [Online] [Cited: 01 10, 2019.] https://www.bloomberg.com/news/articles/2018-04-18/u-s-debt-load-will-be-worse-than-italy-s-by-2023-imf-predicts.

92. [Online] [Cited: 04 09, 2019.] https://www.imf.org/en/Publications/FM/Issues/2018/04/06/fiscal-monitor-april-2018.

93. [Online] [Cited: 04 09, 2019.] https://www.forbes.com/sites/francescoppola/2018/04/17/everything-youve-been-told-about-government-debt-is-wrong/#b2247e0314f4.

94. [Online] [Cited: 01 21, 2019.] https://en.wikipedia.org/wiki/Reserve_currency.

95. [Online] [Cited: 02 05, 2019.] http://www.people-press.org/2017/12/14/public-trust-in-government-1958-2017/.

96. [Online] [Cited: 02 06, 2019.] https://www.bbc.com/news/world-europe-36130006.

97. [Online] [Cited: 02 06, 2019.] http://time.com/time-person-of-the-year-populism/.

98. [Online] [Cited: 06 24, 2019.] https://www.esrb.org/ratings/ratings_guide.aspx.

99. [Online] [Cited: 02 09, 2019.] http://www.apa.org/news/press/releases/2017/03/mental-health-climate.pdf.

100. [Online] [Cited: 02 07, 2019.] https://www.medicalnewstoday.com/articles/322877.php.

101. [Online] [Cited: 02 09, 2019.] https://www.prnewswire.com/news-releases/nearly-three-in-four-millennials-experience-ecoanxiety-300630657.html.

102. [Online] [Cited: 02 09, 2019.] http://headhearthand.org/blog/2018/08/27/just-how-bad-is-the-teen-anxiety-epidemic/.

103. [Online] [Cited: 02 09, 2019.] https://www.tandfonline.com/doi/abs/10.1080/00958964.2011.602131.

104. [Online] [Cited: 02 11, 2019.] https://www.theguardian.com/environment/2018/sep/11/air-pollution-is-biggest-environmental-health-risk-in-europe.

105. [Online] [Cited: 02 11, 2019.] https://www.npr.org/2018/07/10/627687639/estimated-70-deaths-linked-to-canadas-heat-wave.

106. [Online] [Cited: 02 11, 2019.] https://www.theguardian.com/environment/2018/sep/11/air-pollution-is-biggest-environmental-health-risk-in-europe.

107. [Online] [Cited: 02 12, 2019.] https://www.ewg.org/enviroblog/2017/11/study-eating-foods-fewer-pesticides-boosts-women-s-chances-conceive#.Ws4YC9PwYxc.

108. [Online] [Cited: 02 12, 2019.] https://www.cnn.com/2018/04/12/health/chicago-water-lead-report-bn/index.html.

109. [Online] [Cited: 09 30, 2018.] https://www.epa.gov/sites/production/files/2015-05/documents/epa816k02003.pdf.

110. [Online] [Cited: 10 01, 2019.] https://www.livestrong.com/article/196874-long-term-effects-of-chlorinated-water-on-the-skin/.

111. [Online] [Cited: 10 01, 2018.] https://www.cleanwateraction.org/features/dangers-hexavalent-chromium-chromium-6-california-drinking-water.

112. [Online] [Cited: 10 01, 2018.] https://www.ncbi.nlm.nih.gov/pmc/articles/PMC1469238/pdf/envhper00332-0048.pdf.

113. [Online] [Cited: 09 29, 2018.] https://www.jhsph.edu/news/news-releases/2016/study-even-a-little-air-pollution-may-have-long-term-health-effects-on-developing-fetus.html.

114. [Online] [Cited: 09 29, 2018.] https://www.sciencedaily.com/releases/2016/10/161025084744.htm.

115. [Online] [Cited: 09 29, 2018.] http://www.sparetheair.com/health.cfm?page=healthoverall.

116. [Online] [Cited: 09 25, 2018.] http://aqicn.org/scale/.

117. [Online] [Cited: 09 26, 2018.] https://en.wikipedia.org/wiki/Air_Quality_Health_Index_(Canada).

118. [Online] [Cited: 05 23, 2019.] http://blog.filtersnap.com/merv-vs-mpr-vs-fpr-the-definitive-guide/.

119. [Online] [Cited: 09 27, 2018.] https://www.arb.ca.gov/research/indoor/ozone.htm.

120. [Online] [Cited: 09 28, 2018.] http://articles.latimes.com/2008/apr/21/health/he-skeptic21.

121. [Online] [Cited: 09 28, 2018.] https://www.epa.gov/sites/production/files/2018-07/documents/residential_air_cleaners_-_a_technical_summary_3rd_edition.pdf.

122. [Online] [Cited: 10 18, 2018.] https://www.health.harvard.edu/blog/organic-food-no-more-nutritious-than-conventionally-grown-food-201209055264.

123. [Online] [Cited: June 21, 2017.] https://www.washingtonpost.com/news/monkey-cage/wp/2017/05/22/why-its-so-hard-to-know-whether-organic-food-is-really-organic/?utm_term=.33fe84910869.

124. [Online] [Cited: 10 18, 2018.] https://www.usatoday.com/story/life/allthemoms/2018/08/16/organic-foods-contain-lead-and-arsenic-says-report/1009961002/.

125. [Online] [Cited: 10 18, 2018.] https://www.ewg.org/foodnews/dirty-dozen.php.

126. [Online] [Cited: 10 18, 2018.] https://www.ewg.org/foodnews/clean-fifteen.php.

127. [Online] [Cited: 10 18, 2018.] https://www.realsimple.com/food-recipes/shopping-storing/food/non-organic-food.

128. [Online] [Cited: 10 18, 2018.] https://www.salon.com/2015/06/23/18_foods_you_dont_need_to_buy_organic_partner/.

129. [Online] [Cited: 02 12, 2019.] https://www.healthline.com/nutrition/antibiotics-in-your-food#section5.

130. [Online] [Cited: 02 12, 2019.] https://gmwatch.org/en/news/latest-news/18523-glyphosate-found-in-human-hair-as-unique-testing-project-releases-first-results.

131. [Online] [Cited: 02 12, 2019.] https://www.sciencedirect.com/science/article/pii/S0048969717330279.

132. [Online] [Cited: 02 12, 2019.] https://www.healthline.com/nutrition/arsenic-in-rice#section4.

133. [Online] [Cited: 02 12, 2019.] https://www.medicinenet.com/script/main/art.asp?articlekey=212325.

134. [Online] [Cited: 02 12, 2019.] https://www.hsph.harvard.edu/nutritionsource/fish/.

135. [Online] [Cited: 10 21, 2018.] https://www.ewg.org/research/ewgs-good-seafood-guide/executive-summary.

136. [Online] [Cited: 10 21, 2018.] https://www.theglobeandmail.com/life/health-and-fitness/health/whats-the-catch-with-fish/article36839365/.

137. [Online] [Cited: 10 22, 2018.] https://www.scientificamerican.com/article/as-co2-levels-rise-rice-becomes-less-nutritious/.

138. [Online] [Cited: 10 22, 2018.] https://www.npr.org/sections/thesalt/2018/06/19/616098095/as-carbon-dioxide-levels-rise-major-crops-are-losing-nutrients.

139. [Online] [Cited: 10 22, 2018.] https://www.theguardian.com/environment/2014/may/07/climate-change-food-crops-nutrition.

140. [Online] [Cited: 02 12, 2019.] https://www.self.com/story/what-beef-labels-actually-mean.

141. [Online] [Cited: 02 12, 2019.] https://www.farmaid.org/food-labels-explained/.

142. [Online] [Cited: 10 23, 2018.] https://www.cnbc.com/2016/09/09/allergies-are-on-the-rise-and-here-are-three-reasons-why.html.

143. [Online] [Cited: 02 13, 2019.] https://www.ncbi.nlm.nih.gov/pmc/articles/PMC6122668/.

144. [Online] [Cited: 04 11, 2019.] https://www.nbcnews.com/storyline/measles-outbreak/measles-outbreaks-make-2018-near-record-year-u-s-n961276.

145. [Online] [Cited: 10 25, 2018.] https://www.thelocal.fr/20150716/ministry-confirms-700-dead-in-french-heatwave.

146. [Online] [Cited: 02 14, 2019.] https://www.vox.com/2018/7/27/17611940/heat-wave-2018-cost-workers-deaths-health-climate-change.

147. [Online] [Cited: 02 14, 2019.] https://www.globalchange.gov/browse/indicators/us-heat-waves.

148. [Online] [Cited: 10 25, 2018.] https://www.canada.ca/en/health-canada/services/environmental-workplace-health/reports-publications/climate-change-health/communicating-health-risks-extreme-heat-events-toolkit-public-health-emergency-management-officials-health-canada-2011.html.

149. [Online] [Cited: 10 25, 2018.] https://www.weather.gov/ama/heatindex.

150. [Online] [Cited: 12 14, 2019.] https://www.nbcnews.com/health/health-news/here-s-surprising-extreme-heat-risk-1-6-americans-n887566.

151. [Online] [Cited: 10 28, 2018.] http://journals.ametsoc.org/doi/pdf/10.1175/BAMS-D-12-00162.1.

152. [Online] [Cited: 10 28, 2018.] https://en.wikipedia.org/wiki/Polar_vortex.

153. [Online] [Cited: 03 17, 2019.] https://www.charlotteobserver.com/sports/nfl/carolina-panthers/panther-tracks/article217850340.html.

154. [Online] [Cited: 03 17, 2019.] https://ditchingsuburbia.com/blog/remove-ad-clutter.

155. [Online] [Cited: 03 04, 2019.] http://www.neurohackers.com/index.php/fr/menu-top-neurotheque/38-cat-nh-disorders-problems/210-self-esteem-a-materialism.

156. [Online] [Cited: 03 04, 2019.] https://www.psychologytoday.com/ca/blog/in-the-neighborhood/201403/influencing-wellbeing-materialism-and-self-esteem.

157. [Online] [Cited: April 15, 2017.] https://www.canadapost.ca/web/en/kb/details.page?article=how_to_stop_receivin&cattype=kb&cat=receiving&subcat=maildelivery.

158. [Online] [Cited: 12 26, 2018.] https://www.cnbc.com/2018/07/09/paper-straws-are-better-for-the-environment-but-they-will-cost-you.html.

159. [Online] [Cited: 12 30, 2018.] https://www.bloomberg.com/news/articles/2018-10-10/how-much-does-carbon-need-to-cost-somewhere-from-20-to-27-000.

160. [Online] [Cited: 12 30, 2018.] http://www.rff.org/blog/2017/calculating-various-fuel-prices-under-carbon-tax.

161. [Online] [Cited: 12 31, 2018.] https://www.economist.com/graphic-detail/2017/08/29/weather-related-disasters-are-increasing.

162. [Online] [Cited: 12 31, 2018.] https://www.cnn.com/2018/01/08/us/2017-costliest-disasters/index.html.

163. [Online] [Cited: 12 31, 2018.] https://www.usda.gov/oce/foodwaste/faqs.htm.

164. [Online] [Cited: 02 19, 2019.] https://taxfoundation.org/state-pensions-funding-2018/.

165. [Online] [Cited: 02 19, 2019.] https://www.iflscience.com/environment/as-the-climate-changes-and-the-earth-warms-wheres-the-safest-place-on-earth-to-live/all/.

166. [Online] [Cited: 02 19, 2019.] https://www.nytimes.com/2016/10/20/science/9-cities-to-live-in-if-youre-worried-about-climate-change.html.

167. [Online] [Cited: 02 19, 2019.] https://www.businessinsider.com/where-to-live-to-avoid-natural-disaster-climatologists-2018-8.

168. [Online] [Cited: 12 26, 2018.] https://www.statista.com/statistics/678561/urbanization-in-the-united-states/.

169. [Online] [Cited: 12 26, 2018.] https://www.cnbc.com/2018/05/17/two-thirds-of-global-population-will-live-in-cities-by-2050-un-says.html.

170. [Online] [Cited: 12 26, 2018.] https://www.who.int/gho/urban_health/situation_trends/urban_population_growth/en/.

171. [Online] [Cited: 02 19, 2019.] https://www.visualcapitalist.com/mapping-the-worlds-urban-population-in-2050/.

172. [Online] [Cited: 02 19, 2019.] https://www.prb.org/urbanization-an-environmental-force-to-be-reckoned-with/.

173. [Online] [Cited: 02 23, 2019.] https://www.cnbc.com/2017/08/24/most-americans-live-paycheck-to-paycheck.html.

174. [Online] [Cited: 02 23, 2019.] https://www.marketwatch.com/story/lead-poisoning-crisis-sends-flint-real-estate-market-tumbling-2016-02-17.

175. [Online] [Cited: 04 16, 2019.] https://breazy.com/blogs/education/air-quality-and-health-a-guide-to-air-pollution.

176. [Online] [Cited: 04 16, 2019.] https://www.theatlantic.com/business/archive/2015/10/ghost-towns-of-the-21st-century/411343/.

177. [Online] [Cited: 04 16, 2019.] https://www.nar.realtor/sites/default/files/documents/hai-02-2019-housing-affordability-index-2019-04-12.pdf.

178. [Online] [Cited: 04 16, 2016.] https://betterdwelling.com/rbc-canadian-real-estate-affordability-is-at-crisis-levels-but-falling-prices-will-help/.

179. [Online] [Cited: 02 23, 2019.] https://medium.com/datadriveninvestor/why-america-will-devalue-the-dollar-42088956ac18.

180. [Online] [Cited: 04 19, 2019.] https://www.marketwatch.com/story/how-climate-change-could-trigger-the-next-financial-crisis-2019-04-18.

181. [Online] [Cited: 12 12, 2018.] https://www.worldatlas.com/articles/religious-composition-of-the-united-states.html.

182. [Online] [Cited: 12 12, 2018.] https://news.gallup.com/poll/224642/2017-update-americans-religion.aspx.

183. [Online] [Cited: 12 13, 2018.] http://www.worldometers.info/world-population/#religions.

184. [Online] [Cited: 12 13, 2018.] https://www150.statcan.gc.ca/n1/pub/91-003-x/2014001/section03/33-eng.htm.

185. [Online] [Cited: 04 15, 2019.] https://www.mayoclinic.org/healthy-lifestyle/infant-and-toddler-health/in-depth/vaccines/art-20048334.

186. [Online] [Cited: 01 07, 2019.] https://globalnews.ca/news/3973736/flu-shot-fake-news/.

187. [Online] [Cited: 01 07, 2019.] https://www.mayoclinic.org/diseases-conditions/flu/in-depth/flu-shots/art-20048000.

188. [Online] [Cited: 02 24, 2019.] https://www.class-central.com/course/coursera-fake-news-facts-and-alternative-facts-9371.

189. [Online] [Cited: 02 24, 2019.] https://www.class-central.com/course/independent-trust-and-verification-in-an-age-of-misinformation-10629.

190. [Online] [Cited: 03 24, 2019.] https://news.nationalgeographic.com/2016/09/survey-geography-foreign-relations-americans-students/.

191. [Online] [Cited: 03 24, 2019.] https://pressprogress.ca/extra_extra_more_canadians_than_ever_dont_even_follow_the_news/.

192. [Online] [Cited: 04 20, 2019.] https://smallbiztrends.com/2016/03/surepayroll-survey-word-mouth-job-candidates.html.

193. [Online] [Cited: June 15, 2017.] https://en.oxforddictionaries.com.

194. [Online] [Cited: 04 23, 2019.] https://www.eea.europa.eu/data-and-maps/indicators/global-and-european-temperature/global-and-european-temperature-assessment-3.

195. [Online] [Cited: 04 23, 2019.] https://news.nationalgeographic.com/2018/05/plastics-facts-infographics-ocean-pollution/.

196. [Online] [Cited: 04 23, 2019.] https://phys.org/news/2018-12-permafrost-climate.html.

197. [Online] [Cited: 04 23, 2019.] https://www.carbonbrief.org/factcheck-how-global-warming-has-increased-us-wildfires.

198. [Online] [Cited: 04 01, 2019.] https://stilltasty.com/Fooditems/index/19047.

199. [Online] [Cited: 03 15, 2018.] https://www.usda.gov/media/blog/2012/03/22/organic-101-what-usda-organic-label-means.

200. [Online] [Cited: 03 15, 2018.] https://www.npr.org/sections/thesalt/2017/11/02/561462293/hydroponic-veggies-are-taking-over-organic-and-a-move-to-ban-them-fails.

201. [Online] [Cited: 03 15, 2018.] https://www.washingtonpost.com/news/wonk/wp/2017/09/18/lax-enforcement-at-ports-allows-bogus-organic-foods-to-reach-u-s-government-report-says/.

202. [Online] [Cited: 03 15, 2018.] https://en.wikipedia.org/wiki/Galileo%27s_Leaning_Tower_of_Pisa_experiment.

203. [Online] [Cited: 03 15, 2018.] https://earth911.com/business-policy/recycling-plastic-bottles-myths/.

204. [Online] [Cited: 03 15, 2018.] https://www.wikitribune.com/story/2018/03/01/environment/chinas-ban-on-foreign-garbage-creates-mountains-of-waste-in-the-west/49738/.

205. [Online] [Cited: 03 16, 2018.] http://www.postbulletin.com/news/local/mayo-clinic-talks-caffeine-

myths-before-national-coffee-day/article_76ed7182-9fb8-58d9-b08f-64c7e744802c.html.

206. [Online] [Cited: 05 03, 2018.] https://www.scientificamerican.com/article/greenland-is-melting-faster-than-at-any-time-in-the-last-400-years1/.

207. [Online] [Cited: 05 03, 2018.] https://futurism.com/thwaites-glacier-research/.

208. [Online] [Cited: 08 08, 2018.] https://www.ipbes.net/event/ipbes-6-plenary#outcomes.

209. [Online] [Cited: 08 18, 2018.] https://www.bbc.com/news/magazine-35562253.

210. [Online] [Cited: 08 30, 2018.] http://feasibleplanet.com/2018/07/30/getting-back-to-basics/.

211. [Online] [Cited: April 5, 2017.] http://www.gardenmyths.com/garden-myth-born-plants-dont-purify-air/.

212. [Online] [Cited: April 05, 2017.] http://www.buildingecology.com/articles/critical-review-how-well-do-house-plants-perform-as-indoor-air-cleaners/.

213. [Online] [Cited: April 05, 2017.] http://popsop.com/2015/05/hydroponics-at-a-glance-sustainable-alternative-to-soil-gardening/.

214. [Online] [Cited: April 5, 2017.] https://www.washingtonpost.com/lifestyle/food/will-indoor-vertical-farming-help-us-feed-the-planet--or-hurt-it/2016/06/16/f1faaa98-3332-11e6-8ff7-7b6c1998b7a0_story.html?utm_term=.2f65fbdeea14.

215. [Online] [Cited: March 28, 2017.] http://www.marketoracle.co.uk/Article8100.html.

216. [Online] [Cited: 09 06, 2018.] https://www.wikitribune.com/article/76626/.

217. [Online] [Cited: 09 09, 2018.] http://www.npolar.no/en/themes/climate/climate-change/climate-change-in-the-arctic/.

218. [Online] [Cited: 09 27, 2018.] https://www.nafahq.org/understanding-merv/.

219. [Online] [Cited: 10 02, 2018.] http://time.com/4989641/water-air-pollution-deaths/.

220. [Online] [Cited: 10 15, 2018.] https://www.health.com/health/gallery/0,,20599288,00.html.

221. [Online] [Cited: 10 22, 2018.] https://www.health.com/health/gallery/0,,20599288,00.html.

222. [Online] [Cited: 10 22, 2018.] https://www.yallamedi.com/why-choose-antibiotic-free-meat/.

223. [Online] [Cited: 10 23, 2018.] https://news.nationalgeographic.com/2016/04/160408-pollen-climate-change-allergies-spring-seasons/?user.testname=lazyloading:c.

224. [Online] [Cited: 10 23, 2018.] https://www.cdc.gov/mmwr/volumes/67/wr/mm6717e1.htm?s_cid=mm6717e1_x#T1_down.

225. [Online] [Cited: 10 25, 2018.] https://en.wikipedia.org/wiki/Heat_wave.

226. [Online] [Cited: 10 25, 2018.] https://www.thestar.com/news/canada/2018/07/18/89-deaths-now-linked-to-heat-wave-in-quebec.html.

227. [Online] [Cited: 12 26, 2018.] https://www150.statcan.gc.ca/n1/pub/91-520-x/2014001/tbl/tbl2.4-eng.htm.

228. [Online] [Cited: 01 01, 2019.] https://www150.statcan.gc.ca/n1/pub/11-630-x/11-630-x2017004-eng.htm.

229. [Online] [Cited: 01 15, 2019.] https://www.theatlantic.com/business/archive/2011/07/the-us-debt-owed-by-each-american-throughout-history/241366/.

230. [Online] [Cited: 01 23, 2019.] https://www.theguardian.com/us-news/2017/jun/26/jobs-future-automation-robots-skills-creative-health.

231. [Online] [Cited: 01 23, 2019.] https://www.bls.gov/emp/tables/fastest-declining-occupations.htm.

232. [Online] [Cited: 01 23, 2019.] https://www.bls.gov/emp/tables/fastest-growing-occupations.htm.

233. [Online] [Cited: 01 25, 2019.] https://www.forbes.com/sites/gauravsharma/2018/02/20/electric-vehicle-growth-could-see-oil-demand-peak-by-2030/.

234. [Online] [Cited: 01 25, 2019.] https://cleantechnica.com/2017/07/06/electric-cars-will-total-half-new-car-sales-2040-bnef/.

235. [Online] [Cited: 01 25, 2019.] https://energypost.eu/evs-to-make-up-third-of-market-in-2040-e-buses-to-dominate-end-2020s/.

236. [Online] [Cited: 01 29, 2019.] https://www.thehealthyhomeeconomist.com/trihalomethanes/.

237. [Online] [Cited: 02 05, 2019.] http://metrocosm.com/history-of-us-taxes/.

238. [Online] [Cited: 02 09, 2019.] https://www.mnn.com/earth-matters/stories/eco-anxiety.

239. [Online] [Cited: 02 09, 2019.] https://www.livescience.com/55054-kids-mental-health-linked-with-air-quality.html.

240. [Online] [Cited: 02 12, 2019.] https://www.downtoearth.org.in/news/environment/chennai-s-soil-and-delhi-s-air-most-contaminated-due-to-pcb-concentration-study-57217.

241. [Online] [Cited: 02 16, 2019.] https://www.brookings.edu/research/gauging-investment-in-self-driving-cars/.

242. [Online] [Cited: 02 19, 2019.] http://www.citymayors.com/statistics/urban-population-intro.html.

243. [Online] [Cited: April 25, 2017.] https://www.wsj.com/articles/SB113382650575214543.

244. [Online] [Cited: 04 24, 2019.] https://www.nationalgeographic.com/environment/2018/10/microplastics-found-90-percent-table-salt-sea-salt/.

245. [Online] [Cited: 06 26, 2019.] https://www.psychologytoday.com/us/blog/tech-happy-life/201810/what-age-should-i-get-my-child-smartphone.

Index

Other Books by Ken Kroes

Non – Fiction	
 Feasible Planet – A Guide to More Sustainable Living	 Climate Tithing - A Primer
Fiction	
 2022 – Percipience Series Book 1 (Electronic Version is Permafree)	 2222 – Percipience Series Book 2
 2232 – Percipience Series Book 3	 Time Lost - Percipience Series Book 4

Made in the USA
Coppell, TX
05 November 2026

11314460R00184